11th International Conference on Computer and Electrical Engineering (ICCEE 2018)

Journal of Physics: Conference Series Volume 1195

Tokyo, Japan
12 – 14 October 2018

ISBN: 978-1-5108-8940-8
ISSN: 1742-6588

Printed from e-media with permission by:

Curran Associates, Inc.
57 Morehouse Lane
Red Hook, NY 12571

CURRAN ASSOCIATES INC. proceedings.com

Some format issues inherent in the e-media version may also appear in this print version.

Institute of Physics
Dirac House, Temple Back
Bristol BS1 6BE UK

Phone: 44 1 17 929 7481
Fax: 44 1 17 920 0979

techtracking@iop.org

Additional copies of this publication are available from:

Curran Associates, Inc.
57 Morehouse Lane
Red Hook, NY 12571 USA
Phone: 845-758-0400
Fax: 845-758-2633
Email: curran@proceedings.com
Web: www.proceedings.com

TABLE OF CONTENTS

ELECTRONICS AND COMMUNICATION ENGINEERING

POWER MACHINERY & MEASUREMENT AND CONTROL TECHNOLOGY

ELECTRONIC POWER TECHNOLOGY AND ENERGY

IOP Conf. Series: Journal of Physics: Conf. Series **1195** (2019) 011001 doi:10.1088/1742-6596/1195/1/011001

Preface

We are very pleased to welcome all of you to 11th International Conference on Computer and Electrical Engineering held in Tokyo, Japan during October 12-14, 2018. ICCEE was started in Phuket Island, Thailand in the year of 2008 and after the success of the first edition, it has been held annually from 2009 to 2017 in Dubai (UAE), Chengdu (China), Singapore, Hong Kong, Paris (France), Geneva (Switzerland), Paris (France),Barcelona (Spain), and Edmonton (Canada). With the successful experience over the past 10 years, This year, ICCEE starts off for the new decade.

The goal of ICCEE2018 is to discuss the latest research and results of scientists and engineers from academic side and industrial side related to computer and electrical engineering topics. All the papers were subjected to peer-review by conference committee members and international reviewers. We had 55 submissions and accepted 30 high quality papers related to computer and electrical engineering such as data science, software engineering, image analysis, computer science, control technology, electronic power technology and so on. The acceptance rate was 54.5%. The attendees are from various regions such as Asia, Oceania, Europe and Africa. We also have five invited speakers from US, India and Japan. The proceedings give an exciting and wide-ranging discussion of the topics presented at ICCEE2018.

The ICCEE2018 has been organized in the program chapters as: Data Science and Software Engineering; Image analysis and processing technology; Computer Science and Information Engineering; Electronics and Communication Engineering; Power machinery & measurement and control technology; Electronic power technology and energy.

The ICCEE2018 peer-reviewed and accepted papers have been edited as conference proceedings to be published with Journal of Physics: Conference Series (doi:10.1088/issn.1742-6596; Online ISSN: 1742-6596 / Print ISSN: 1742-6588), which is indexed by Conference Proceedings Citation Index – Science(CPCI-S) (Thomson Reuters, Web of Science),Scopus, Ei Compendex, Inspec(IET).

In the end, we appreciate all the authors who have contributed to this conference and also to the reviewers, speakers, chair persons and all the conference participants for

their support to ICCEE 2018. We are also very grateful to the many people who helped with the organization of the conference. Hope to see you in ICCEE2019 again!

Best Regards.

Assoc. Prof. Dr. Mitsuharu Matsumoto
Program Chair of ICCEE2018
University of electro-communications, Japan

COMMITTEES

Conference Chairs
Prof. Peter Plapper, University of Luxembourg, Luxembourg
Prof. Genci Capi, Hosei University, Japan

Program Chairs
Assoc. Prof. Dr. Matsumoto Mitsuharu, University of electro-communications, Japan
Prof.Viacheslav Pshikhopov, Southern Federal University, Russian Federation

Steering Committee
Prof. Sathans Suhag, National Institute of Technology Kurukshetra, India
Prof. Takashi Harada, Kindai University, Japan
Asst. Prof. Dr. Rafiq Ahmad, University of Alberta, Canada

Publicity Chair
Asst. Prof. Hassan HosseinNia, TU Delft/3mE/Precision and Microsystems Engineering (PME), Netherlands

Technical Committee
Prof. Basel Alsayyed Ahmad, University of Alberta, Canada
Asst. Prof. Dr. Hossein Rouhani, University of Alberta, Canada
Asst. Prof. Dr. Ahmed Qureshi, University of Alberta, Canada
Dr. Sattar J Aboud, Science and Technology of the Iraqi Council of Representatives, Iraq
Dr. Ugur Bilge, Akdeniz Üniversity Faculty of Medicine, Turkey
Prof. Said Yahmedi, Laboratory of Mathematical Modeling and Numerical Simulation Annaba, Algeria
Prof. Kamel Eddine HEMSAS, University Ferhat Abbas Setif -1-, Algeria
Prof. Dr.Ing. Rainer Müller, Zentrum für Mechatronik und Automatisierungstechnik, Germany
Dr.Earle Jennings, QSigma, Inc., US
Asst. Prof. Dr. Waheeb Abu-Dawwas, Applied Science Private University, Jordan
Asst. Prof. Dr. Levent Seyfi, Selcuk University , Turkey
Prof. T. Cetin Akinci, Istanbul Technical University, Turkey
Asst. Prof. Dr. Akif Durdu, Selcuk University, Turkey
Prof. Parvin Lakbala, Hormozgan University of Medical Sciences, Iran
Asst. Prof. Dr. Ali Hassan, National University of Sciences and Technology, Pakistan
Asst. Prof. Dr. Claude Tadonki, Mines ParisTech - CRI, France
Dr. Tsung-Mou Huang, Taiwan Power Co. Taiwan
Prof. Samuel Xavier-De-Souza, Universidade Federal do Rio Grande do Norte , Brazil
Dr. Kemal UCAK, Muğla Sıtkı Koçman University, Turkey
Assoc. Prof. Dr. Vasudev Malhotra,Y.M.C.A University of Science and Technology, India
Asst. Prof. Dr Abbas Amini, Western Sydney University and Australian College of ACK, Australia
Dr. Boris Gurenko, Southern Federal University, Russia
Dr. HASSAN ELAHI, Institute of Space Technology, Pakistan

IOP Conf. Series: Journal of Physics: Conf. Series **1195** (2019) 011003 doi:10.1088/1742-6596/1195/1/011003

Peer review statement

All papers published in this volume of *Journal of Physics: Conference Series* have been peer reviewed through processes administered by the proceedings Editors. Reviews were conducted by expert referees to the professional and scientific standards expected of a proceedings journal published by IOP Publishing.

Pattern recognition of UAV flight data based on semi-supervised clustering

N Wang[1,3,4,5], Z SH Xu[2], SH W Sun[1,3], Y Liu[1,3,4,5]

[1]School of Management Science and Information Engineering, Jilin University of Finance and Economics,China
[2]Shenyang Woozoom Technology Co., Ltd,China
[3] Laboratory of Logistics Industry Economy and Intelligent Logistics, Jilin University of Finance and Economics,China
[4]Jilin Big Data Research Center for Business, Jilin University of Finance and Economics,China
[5] Jilin Province Key Laboratory of Fintech, Jilin University of Finance and Economics,China

Email: ctuwangnan@126.com

Abstract. UAV is an unmanned aerial vehicle controlled by a remote radio signal or a trajectory planning software carried itself. It is widely used in military, civil and scientific research fields. However, due to the lack of real-time decision-making ability, the UAV has high fault rate. The flight quality assessment of UAV and the construction of fault prediction model can be used for debugging and fault-removing to customer's aircraft, and also to increase the added value of the civilian UAV products. Before building a fault prediction model, a very important step is to identify the pattern of sampled data. For each group of flight data, the efficiency and accuracy rate of manual quality evaluation and fault identification are not acceptable. Based on the UAV flight data accumulated in the big data platform of an UAV production company in Shenyang, Liaoning Province of China, this paper proposes a semi-supervised clustering technique to do automatic pattern recognition for the sampling points. According to the characteristics of UAV flight data, two different methods are designed to choose initial centroids. Meanwhile, we use the existing normal flight data to train distance thresholds to combine some clusters to eliminate the resulting error clustering. Real flight data or flight test data with manually added labels are used to run the proposed algorithms to verify the recognition results. The experimental results show that the proposed methods greatly improve the efficiency and accuracy of adding precise labels to the historical flight data and play a role in assisting the manual recognition of sampling points while strengthening the management and statistics.

1. Introduction

The unmanned aerial vehicle (UAV) has a significant improvement in its autonomous ability compared with the manned vehicle, and the ability to survive is greatly strengthened. The UAV can replace humans to complete the task in a bad environment and do not need to worry about the safety of the driver's life, and can perform the task of high risk[1,2,3,4]. So, UAV has been widely applied in military, civilian and scientific research fields. The widely application of UAV requires that it should have a high stability in addition to the efficient completion of the task, that is, the UAV is required to

detect the health status of itself and take the corresponding scheme according to the current state[5]. Due to the lack of real-time decision-making ability, the UAV has high fault rate. Fault prediction is the core of UAV health management technology. The flight quality assessment of UAV and the construction of fault prediction model can be used for debugging and fault-removing to customer's aircraft, and also to improve the added value of the civilian UAV products. The methods of fault diagnosis can be divided into model-based and model-free methods. When the model is not accurate, the diagnostic accuracy of the model-based diagnosis method is low, so the diagnosis method that does not rely on the model is getting more and more attention. The advantage of the model-free method is that it does not require the accurate model of the system and has strong self-adaption ability. Its defects are complex construction making it not conducive to practical application [6]. The model-free methods can be classified into data driven, knowledge based and discrete event-based methods. UAV produces a large number of real-time data during the flight process. Using these data and machine learning algorithms, such as neural network-based methods [7,8,9,10], the fault prediction model of aircrafts can be constructed. However, the real-time data produced in the flight process are all unlabelled data, which is not to determine which sampling points in a flight data are fault-free points of normal flight and which are the fault points. So before building a fault prediction model, a very important step is to identify the pattern of the sampling data, which is also partially for improving the whole flight portrait. The interval of sampling points for UAV flight data is usually in milliseconds, for example, the data involved in this paper comes from the big data platform of a Shenyang UAV production company, with the data sampling interval of 2ms. For each group of flight data, the efficiency and accuracy rate of manual quality evaluation and fault identification are not acceptable.
All measurement variables have strong correlations, rather than independent characteristics. Therefore, we speculate that the pattern between the variables in normal flight will be different from that after fault happens. Running the proposed clustering methods on one group of flight data, the patterns of the sampling points are distinguished. In this algorithm, two different methods of choosing initial centroids are designed according to the characteristics of UAV flight data. Meanwhile, we use the existing normal flight data to train distance thresholds to combine some clusters to eliminate the resulting error clustering. We use real flight data or flight test data to test our methods. The experimental results show that the proposed methods greatly improve the efficiency and accuracy of adding precise labels to the historical flight data and play a role in assisting the manual recognition of sampling points while strengthening the management and statistics.

2. Data pre-processing

2.1. Feature selection

The UAV system consists of the controlled object (aircraft), the controller (generally including the measuring element, the sensor) and the actuator (steering engine, propeller). In addition, some other auxiliary positioning components (such as millimeter wave radar) may also be installed on the basis of the specific circumstances. The flight state of the aircraft is automatically determined by the flight control system according to the height or vertical speed of the aircraft. The flight mode of the aircraft can be controlled by the remote controller. The control volume of the remote controller channel is different under different flight modes. According to the flight mode and control volume or certain mission planning, the flight control system calculates the expectation data of location and speed. The controller calculates some control quantities based on these control expectations and current aircraft status and drives the actuator to control the behaviour of the important components, such as accelerometers, gyroscopes, magnetic compass and other ones, so that the aircraft will produce a certain mode of motion. The flight sensors mounted on these components measure these movements, including the pitch angle, the roll angle, the corresponding angular velocity and other variables, and return the calculated values, and the barometer returns the measured height. These values as actual values are compared with expectation values, and some control quantities for adjusting attitude are obtained, and then, the flight is corrected. Therefore, we can filter out the data generated during the

flight process, remove the intermediate variables in the calculation process, and retain the important feature variables.

The UAV flight control system adopts closed loop control. In different flight modes, the variables involved in the control and calculation are different. There are more than one flight mode switching in one flight data. Therefore, the unified characteristic variables cannot be used. For example, for multi axis UAV, in attitude mode, the index of speed is not involved in calculation and control. For the helicopter's unique mode of non-aileron, the indices of attitude, position and speed can be eliminated. Therefore, this paper runs clustering algorithm respectively on different flight modes.

2.2. Data normalization

After manual screening, we have retained a total of 51 characteristic variables for the flight data obtained by the big data platform, all of which are numerical data. Because the range gap between the feature variables is large and the value of each characteristic variable tends to be concentrated, the linear normalization function (Min-Max Scaling) is adopted to convert the original data value to the value in range of [-1,1]:

2.3. Feature merging and feature weighting

2.3.1. Feature merging. Abrupt separation of three axis posture is one common fault which will produce the poor following values of Pitch, Roll and Yaw. According to this, we get the new combination features through the difference between the expectation value and the measure value of Pitch, Roll and Yaw.

2.3.2. Feature weighting. Another common fault is sensor failure. Data show that about 50% of UAV faults are caused by sensor failures [11]. When the fault is identified manually, it is found that the three-axis data change returned by the accelerometer is the most dominant fault data, which generally indicates that the accelerometer vibration suddenly becomes larger and is the most frequent fault type. Data normalization eliminates the influence of eigenvalues on distance calculation results, thus weakening the change of this mode. We dynamically weighted two sets of three axis data returned by the accelerometer (one set is the fixed data, the other is the shock absorption data), that is, the greater the three axis values of a certain sampling point being higher than its previous sampling point, the higher the weight value added to it.

The weight function is $x_x\ x\ xxxxxxxxx_x x\ xxx_{xxx}x$, in which, '$xxxx$' represents Euclidean distance function, xxx_x is the vector composed of the two sets of three axis data. If there is a fault, the acceleration value will change greatly, and the corresponding weight of the point will be bigger. However, this does not distinguish between a real fault and a sharp change in the acceleration caused by the unsuitable operation of human operator, so that these will also be classified as fault data.

3. Algorithm design

3.1. Choose the number of clusters

For simplification, we only consider one kind of fault in flight process, or a most significant fault mode, not considering the mode of multiple fault superpositions for now. In fact, through the analysis of the flight data, we find that in the actual flight process, there are few fault superimpositions, even if there is, it is also such a situation that a component fault happens first, and before a short time of crashing, or in the moment of crashing, there may appear several other faults. After fault happens, there are two cases of crashing and non-crashing for the aircraft. We hope to identify the normal flight data points, the fault points (if crashing, also containing the crashing point), and the points after crashing (if crashing). Therefore, the selected number of clusters is 2 (not crashing) and 3 (crashing

IOP Conf. Series: Journal of Physics: Conf. Series **1195** (2019) 012001 doi:10.1088/1742-6596/1195/1/012001

after faults happen), and the clusters and the number of clusters with the minimum error are output as the final clustering results.

For the flight data without fault, we pre-train cluster centroids distance threshold, and two clusters with the inter centroid distance less than the threshold are considered to be an error pattern recognition. For example, for a group of flight data without fault, all of the sampling points in theory should belong to the same cluster, but because the minimum clustering number is 2, the clustering results are not reasonable. At this time, the distance between the two cluster centroids is smaller than that of the fault-free points cluster and the fault points cluster and so, we will merge the two clusters.

3.2. Choose the initial centroids

We design two methods to choose the initial centroids and run the proposed clustering algorithm respectively:
(1) Random selection of *k* sampling points as the initial centroids;
(2) The two adjacent sampling points with the maximum change amplitude of the three axis acceleration values are selected as the first two initial centroids, and the farthest sampling point from the set of the first two initial centroids is selected as the third centroid.

The third centroid in the method (2) determines p^j which is farthest from set *S* (here composed of the first two initial centroids) by solving the following optimization problems, namely the optimal function:

$$max_{p^j \in P-S} min_{p^i \in S} D(p^j, p^i),$$

This optimal function represents: For each points p^t （$1 \le t \le |P - S|$, $|P - S|$ represents the number of the points in set $P - S$）, let d_t be the smallest distance from p^t to all points of *S*, a^j is the farthest point from *S* if $d_j = max_{1 \le t \le |A-S|}\{d_t\}$.

3.3. Flight data pattern discrimination algorithm based on k-means clustering

For the method of randomly selecting *k* centroids, we run *k*-means clustering algorithm 10 times on each group of data and get the clustering result with the smallest clustering error. The clustering error is computed by SSE (Sum of Squared Errors), and the distance between sampling points is calculated by Euclidean distance formula.

The flight data pattern discrimination algorithm based on *k*-means clustering is described as PattDiscrBasedonKMeans, in which we suppose there are *m* kinds of flight modes.

Algorithm. PattDiscrBasedonKMeans(DataSet, GenerateCentMethod, ω):

    # DataSet is the data set to be clustered, GenerateCentMethod is the two proposed methods to get the initial centroids, and ω is the centroid distance threshold of merging fault-free data.

    Do feature merging and feature weighting for DataSet

    Choose feature variables for DataSet according to the flight mode: $DataSet_1$~ $DataSet_m$

    For *k* in {2,3}, do:

        SSE_k=0

        For each $DataSet_i$($1 \le i \le m$):

            $originalCentroids_i$=GenerateCentMethod(DataSet, *k*)

            $centroids_i$, $clusterAssment_i$, SSE_i=kMeans($DataSet_i$, $originalCentroids_i$)

            # $clusterAssment_i$ is the classifying results for each point of $DataSet_i$, SSE_i is the sum of squared errors for all points, $centroids_i$ is the set of centroids of the resulted clusters

            # Here if the method of randomly selecting *k* centroids is adopted, it will output the clustering result with the smallest clustering error of running 10 times *k*-means clustering algorithm

            Determine the two points in $centroids_i$ with the smallest distance, if the distance is smaller than ω, merge the corresponding clusters, derive the centroid of the new cluster

and revise the value of SSE_i. Repeat this step until all points are merged to one cluster or the distance is not smaller than ω.

$SSE_k=SSE_k+SSE_i$

Number the m group of clustering results separately, combine them as the final clustering result of DataSet

Output the clustering results corresponding to the smaller values of SSE_2 and SSE_3

4. Experimental verification

4.1. Choosing data and evaluation index

4.1.1 Choosing experimental data. The experimental data of this paper come from the real flight data of the big data platform of an unmanned aerial vehicle (UAV) company in Shenyang City, Liaoning Province of China. After screening the characteristics of the original data, we chose 17 sets of fault flight data (199883 sampling points) and 29 groups of normal flight data (450996 sampling points).

The original flight data dimension obtained by the big data platform is very high, which contains all kinds of meaningless intermediate calculation results. We discussed with the company's technical employees involved in our research to select 51 eigenvalues at last. After choosing proper flight data and manually identifying the fault from large flight records, 17 sets of fault flight data were obtained. These data are all with manual labels, including the identification of normal points, fault occurrence point (if crashing happened, also including the crashing point), the points after crashing (if crashing happened). At the same time, it also includes the manual discrimination about the cause of the fault and the reason for crashing.

Table 1. Seventeen groups of fault flight data.

Data No	Number of flight modes	Number of all sampling points	Number of predefined normal points	Number of predefined fault points(including crashing point)	Number of predefined after-crashing points
1	3	13695	9204	4479	12
2	4	6244	4911	984	349
3	2	21535	17849	87	3599
4	2	1687	776	911	0
5	2	8034	7665	347	22
6	2	23601	21875	1715	11
7	2	5534	4291	121	1122
8	2	2815	2676	118	21
9	4	13681	11656	1949	76
10	1	9296	9200	96	0
11	3	9399	8688	97	614
12	2	9416	9261	155	0
13	1	3841	3743	98	0
14	1	8442	7616	394	432
15	1	12262	12156	106	0
16	2	28055	25270	2785	0
17	2	22346	9783	4418	8145

4.1.2. Evaluation. The common clustering evaluation indexes are purity and F-value, and F-value is more commonly used. The calculation of F includes two indexes: Recall rate and Precision rate.

- precision[*i*][*j*]=the number of those points predefined as the *i*th class and classified to the *j*th cluster/the number of points in the predefined *j*th class
- recall[*i*][*j*]=the number of those points predefined as the *i*th class and classified to the *j*th cluster/the number of points in the predefined *i*th class
- f[*i*][*j*]=2*precision[*i*][*j*]*recall[*i*][*j*] / (precision[*i*][*j*]+recall[*i*][*j*])

In this way, we get the F values between each predefined class and each derived cluster. We generate the weighted mean value of F obtained by the different initial centroids selection methods used in this paper.

F-measure=sum(f[*i*][*j*]*the number of points in the predefined *i*th class) / the number of total points

The larger the value is, the better the clustering result is.

4.1.3. Experimental design. In this paper, we do experiments to 17 groups of fault flight data respectively. Two different methods are used to select the initial centroids, and the corresponding F values are calculated. At the same time, the accuracy of discrimination for the fault occurrence point is calculated.

In the course of the experiment, in accordance with the different flight mode, 29 groups of normal flight data are used to run the clustering algorithm with cluster number of 2. The distance between the two clusters of each flight mode is calculated, and the average value of the distances of the 29 sets of data is taken as the threshold for merging the clusters of the normal sampling points.

4.2. Results and analysis

For each group of flight data, we cluster it according to different modes, and then combine and manually correct the clustering results of different flight modes according to the time stamp of the sampling points.

In the clustering results obtained in a certain mode, the cluster composed of those points closer to starting time is classified as the normal class(fault free); the cluster composed of those points closer to ending time is classified as the after-crashing class (if the results are 3 clusters); the other points are classified as the fault class.

Thus, we get the final clustering result of a group of flight data. Then we calculate the values of indexes for the clustering results when running the clustering algorithms with different methods of generating initial centroids: Method1 and Method2, corresponding to the two methods in Section 3.2. The results are shown in Table 2 and Table 3.

Table 2. The values of indexes for the clustering results of Method1.

Data No	Number of classified normal points	Number of classified fault points	Number of classified after-crashing points	Number of generated clusters	Predefined number of clusters	F-Value	The recognition rate of fault points
1	9189	4506	0	2	3	0.89	87.65%
2	4877	1077	290	3	3	0.88	84.75%
3	17768	299	3468	3	3	0.90	86.66%
4	767	920	0	2	2	0.84	83.51%
5	7767	267	0	2	3	0.88	70.14%
6	21878	1723	0	2	3	0.89	85.54%
7	4317	207	1010	3	3	0.88	85.01%
8	2689	126	0	2	3	0.85	39.58%

9	11436	2148	97	3	3	0.89	86.50%
10	9168	128	0	2	2	0.89	36.92%
11	8705	98	596	3	3	0.89	79.37%
12	9267	149	0	2	2	0.89	56.74%
13	3679	162	0	2	2	0.88	37.98%
14	7879	83	480	3	3	0.89	81.92%
15	12101	161	0	2	2	0.89	41.83%
16	25988	2067	0	2	2	0.90	87.20%
17	9689	3656	9001	3	3	0.89	87.30%
Avg						0.88	71.68%

Table 3. The values of indexes for the clustering results of Method2.

Data No	Number of classified normal points	Number of classified fault points	Number of classified after-crashing points	Number of generated clusters	Predefined number of clusters	F-Value	The recognition rate of fault points
1	9200	4495	0	2	3	0.90	90.26%
2	4967	987	290	3	3	0.88	87.83%
3	17878	289	3368	3	3	0.90	89.23%
4	767	920	0	2	2	0.88	87.50%
5	7731	287	16	3	3	0.87	85.54%
6	21879	1703	19	3	3	0.90	90.21%
7	4274	150	1110	3	3	0.90	89.18%
8	2695	120	0	2	3	0.88	75.13%
9	11438	2148	95	3	3	0.90	89.50%
10	9168	128	0	2	2	0.88	87.50%
11	8696	107	596	3	3	0.87	87.04%
12	9200	216	0	2	2	0.91	90.50%
13	3679	162	0	2	2	0.91	90.50%
14	7677	315	450	3	3	0.88	87.73%
15	12111	151	0	2	2	0.90	89.50%
16	25765	2290	0	2	2	0.91	90.50%
17	9691	3655	9000	3	3	0.88	87.73%
Avg						0.89	87.96%

The average recognition rates of the two methods to the fault points are 71.68% and 87.96%, respectively. But we find that both of the two methods have low recognition rates for the fault starting point and the crashing point. Only by using the second method, in two of the 17 sets of data the fault starting point are relatively accurately located (the error is 2 sampling points and 13 sampling points). The two methods have no more than 200 sampling points (0.4 seconds) for the forward or backward errors in the fault location, but in practical applications, this error is still somewhat large.

The F values obtained by the two methods are both relatively large, which indicates that the effect of clustering is good, and it is found that most of the sampling points in consecutive time are in the same cluster. If after the automatic clustering, a certain manual correction is added, for example, to make the continuous sampling points belong to the same cluster, the recognition rate of the fault points will be greatly improved.

5. Conclusion

The big data platform of UAV enterprises accumulates a lot of real time data during the flight process of UAV. Using these data to warn the fault of aircraft during flight and timely adopting corresponding measures can reduce the loss for users caused by the crashing.

The first step to establish the early warning mechanism is to identify the types of faults that may happen in the aircraft, and to label the large number of historical flight data accurately.

In this paper, a semi-supervised clustering technique is used to distinguish normal sampling points and fault sampling points in a flight process, which greatly saves the time of manual recognition and improves the accuracy of recognition. We will further study from the following two aspects in future: (1) By improving the initial centroids selection, merging features and modifying distance formula, we can improve the clustering algorithm and further improve the accuracy of fault recognition. (2) Merging multiple groups of flight data of the same type of aircraft, finding the optimal number of clusters and identifying a variety of fault types.

Acknowledgments

This work is supported in part by NSFC under Grant No. 61702213, Jilin Provincial Department of Education "Thirteenth Five Year Plan" Science and Technology Development Plan under grant No.JJKH20180463KJ, Jilin Province Science and Technology Development Plan under Grant No.20180101337JC, the Open Project of Laboratory of Logistics Industry Economy and Intelligent Logistics under Grant No.201701.

References

[1] Wang H B, Ge X Y and Song Y Zh 2009 The development trend of UAV and forecast technology *Winged Missiles Journal* 11 46–52.

[2] Zou X F, He Q H, He J L 2006 Development status and related technologies of UAV *Winged Missiles Journal* 10 9–14.

[3] Liu Ch Y 2010 Development of UAV technology in foreign countries *Ship Electronic Engineering* 30(1) 19–23.

[4] Wallace L, Lucieer A and Watson C 2012 Development of a UAV-Li DAR System with Application to Forest Inventory *Remote Sensing* 4(6) 1519–1543.

[5] He M L 2015 Researches on fault diagnosis and fault-tolerant control for unmanned aerial vehicle *Master Thesis of Guizhou University*

[6] Shen Y, Ding S X and Haghani A 2012 A comparison study of basic data-driven fault diagnosis and process monitoring methods on the benchmark Tennessee Eastman process *Journal of Process Control* 22(9) 1567–1581.

[7] Hao T, Tang Y Zh, Ren Y Q 2008 Application of BP neural network in sensor fault diagnosis for flight control system *Computer Measurement & Control* 16(5) 613–615.

[8] Li M, Xu X D 2002 Powell neural network method for sensor fault detection *Thermal and Power Engineering* 17(1) 73-75.

[9] Gu L CH, Zhang Y Y, Qiu D 2002 Research on Sensor Failure Detection and Diagnosis Based on Neural Network Journal of xi'an jiao tong university 36(9) 959-962.

[10] Wang J P 2002 Fault diagnosis technology based on Neural Network Multisensor data fusion *Mechanical science and technology* 21(1) 127-130.

[11] Zeghlache S, Saigaa D Kara K 2016 Fault tolerant control based on neural network interval type-2 fuzzy sliding mode controller for octorotor UAV *Frontiers of Computer Science* 10(4) 657-672.

Cloud storage platform for efficient RDF compression

Y X Sun[1], S H Lee[2] and Y J Lee[3]

[1,2,3]School of Computer Science and Engineering, Kyungpook National University, Daegu, Korea.
syx921120@gmail.com

Abstract. In current web environment, a large of RDF datasets are producing. Thus, the storage of RDF data is becoming an important part of Semantic Web development. Due to the triple structure of RDF data, how to efficiently compress RDF data as much as possible without breaking the integrity of data is become one of the important issues. by Comparing previous research, we found that people had made many efforts in the compression of static, large RDF datasets. In this paper, we not only propose a cloud-based compression approach based on previous research, but also made efforts in data security. Due to the openness of the internet, anyone can publish their Linked Data. At this time, the security of personal data will become very important. Our cloud storage platform not only provides publishers with efficient compression services but also maximizes the security of data. To improve the openness of the platform, we adopt REST (REpresentational State Transfer) architecture to exchange data between publishers and cloud.

1. Introduction

The principles of Linked Data were first outlined by Berners-Lee in 2006 [1]. Linked Data is about utilizing the Resource Description Framework (RDF) and the Hypertext Transfer Protocol (HTTP) to publish structured data on the Web and to connect data between different datasets. In LOD (Linked Open Data) cloud [2], From 2007 to 2017, the number of datasets had increased by nearly 100 times. Although the growth of datasets is beneficial for the development of semantic web, we must face the storage pressure of RDF data. In [3], three basic RDF compression approaches are proposed. By analysing the experimental results, we find that RDF data is highly compressible. In [4] a binary RDF representation was proposed, because it was consisted by Header information, a Dictionary, and the actual Triples structure, thus it was named HDT. It is mainly about how to improve high levels of verbosity/redundancy and weak machine-processable capabilities in the description of these datasets. according to the authors, it outperforms existing compression solutions for efficient RDF exchange. In [5] authors proposed a logical Linked Data compression approach. It is a novel lossless compression technique for RDF datasets, called Rule Based Compression (RB Compression) that compresses datasets by generating a set of new logical rules from the dataset and removing triples that can be inferred from these rules. It can prune more than 50% of the original triples without affecting data integrity. Although these approaches can compress RDF datasets efficiently, their datasets have a common point is large and static. The user-generated datasets have an important feature is dynamic and need to pay more attention to data security. Thus, for addressing these issues, we proposed a dictional-based cloud storage platform. Not only we can reduce local storage pressure but also improve the security of published data. Users don't need to access local server to modify published data, just only by REST-APIs[6] to modify dictionary data in the cloud platform.

IOP Conf. Series: Journal of Physics: Conf. Series **1195** (2019) 012002 doi:10.1088/1742-6596/1195/1/012002

In related work, we will summarize these compression approach's feature. Later, we will propose cloud storage platform's system architecture and service. In experiment part, by experiment analysis we can obtain actual compression performant. Finally, in conclusion part we will make an evaluation about cloud storage platform.

2. Related work

Table 1 shows three basic compression approaches in [3]. Direct Compression uses three well-known techniques such as gzip, bzip2, and ppmdi. Adjacency Lists focus the data repeatability. For example, the set of triples can be converted into the adjacency list s → [(p_1, $ObjList_1$), ⋯ (p_k, $ObjList_k$)]. Dictionary+Triples split the data into dictionary of elements and the triples substituting for each element. Triples can be represented by adjacency lists. In order to obtain better reflect the influence of data structure on the compression results, we adopt the same compression approach to compress data of different structures. Here, gzip approach was adopted.

Table 1. Basic compression approaches.

Approach	Description
Direct Compression	Compression with gzip, bzip2, ppmdi
Adjacency Lists	Convert RDF data to adjacency lists
Dictionary + Triples	Split data into dictionary and triples

In figure.1, Direct Compression, Adjacency Lists, and Dictionary+Triples compressions refer to as DC, AL, and DT. We analyze the three basic approaches and test them with a well-known real world RDF data set, DBpedia[7]. From this test, we found that the using of AL greatly reduce storage space. DT shows slightly worse performance than other two approaches. The main reason of the DT low compression is dictionary itself. Nevertheless, we decide to adopt DT, because DT is focused on the graph nature of RDF. To reduce the dictionary storage pressure, therefore, we implement a cloud storage platform for the dictionary management. We could achieve the best storage performance by separating the dictionary and adjacency lists. This conclusion is also verified in [3]. This provides a powerful theoretical basis for the development of our cloud storage platform

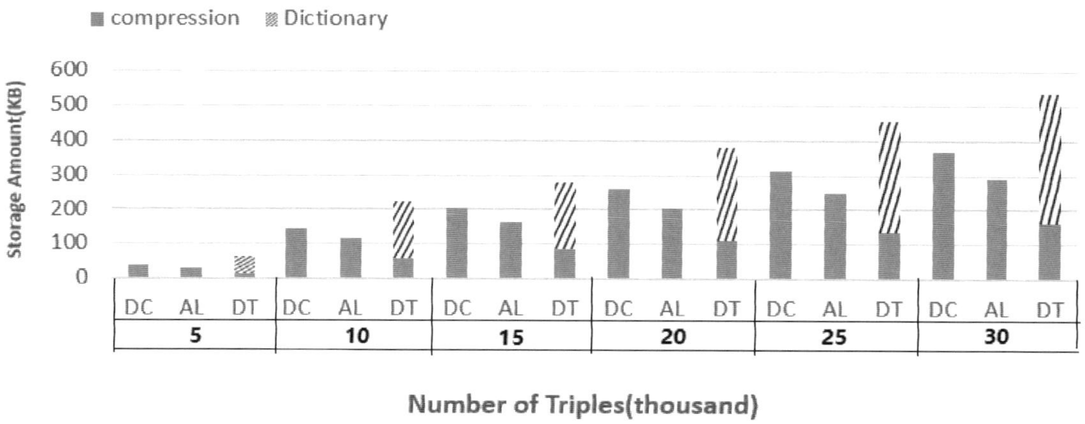

Figure 1. Performance of different compression approaches.

In [8] Because each RDF statement is made of three different terms: a subject, a predicate, and an object. Thus, authors proposed a scalable compression approach for large RDF datasets by using the MapReduce programming model [9] and dictionary encoding technique. According author's description this technique can be used by all RDF applications that need to efficiently process a large amount of data, such as RDF storage engines, network analysis tools, and reasoners.

The aforementioned approaches are focus on large, static RDF databases, some of these technologies are very helpful for the development of our cloud storage platform. In this paper, we not only define a lossless compression approach but also the dynamic modification and data security can be realized.

3. System architecture

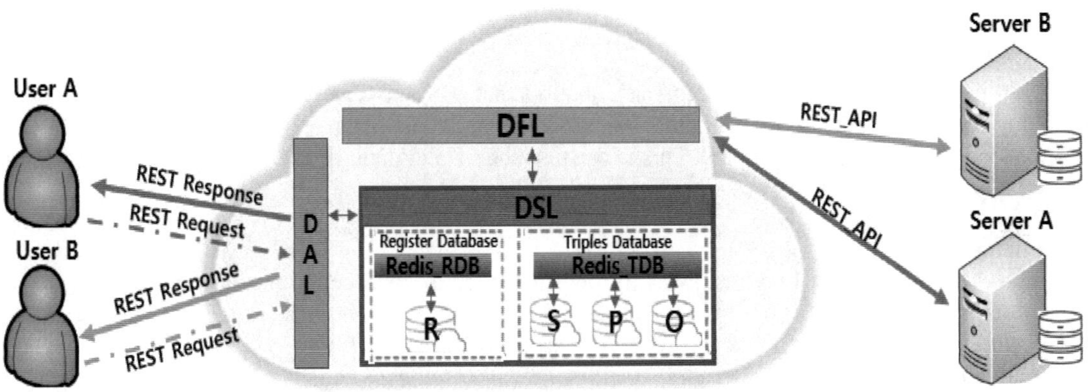

Figure 2. Architecture of cloud storage platform.

Figure.2 shows the architecture of the cloud storage platform. The cloud platform is divided into three layers, DAL (Data Analysis Layer), DFL (Data Forward Layer) and DSL (Data Storage Layer). DSL is divided into a register database and a triples database. The register database mainly stores user's ID and forward URL for target server's REST-API. The triples database mainly stores hash values and string values of triples and they are independent of each other. Notice, server A and server B are not part of the cloud platform. It is user's personal server, that's why we set up the register database. Therefore, users can change the access path of the server at any time by modifying the information of the register database. The following explains two main services for the cloud storage platform.

Data Parsing and *Registration Service*: Registration must be completed before users upload original RDF datasets. In DSL, the cloud platform will parse the original data and stores it in a triples database. In the meantime, a compression file is generated. In this case, the compressed file is an adjacency list based on hash values. Users can download this compression file by REST-API and stores it in local server.

Data Query Service: If users want to query data that they only need to access the cloud platform, in register database they will obtain the URL of target server, then they can obtain result by DFL. At this point, the result is an array of hash values. If users want to obtain final results they need to convert the hash value into a real value in the triples database. Complete above steps, users can get the final results that they want.

4. Experiment

For making the results more convincing, we downloaded datasets from three different fields. The DBpedia dataset contains RDF information extracted from Wikipedia. The DrugBank dataset contains information on drugs and drug targets. Table 3 gives the characteristics of the collected dataset. It is

widely used by the drug industry, medicinal chemists, pharmacists, physicians, students and the public institutions. The LinkedGeoData dataset is a large spatial knowledge base which has been derived from OpenStreetMap for the Semantic Web. table 2 shows detailed performance of compression, our adjacency list structure uniformly adopts s → [(p1, ObjList1), ··· (pk, ObjListk)].

Table 2. Result of cloud storage platform compression.

	DBpedia	DrugBank	LinkedGeoData
Original size(MB)	3.94	144	327
Number of triples	31,050	766,920	2,207,295
Compressed size(KB)	173	2,821	12,182
Compression rate	4.29%	1.91%	3.63%

From the results of the experiment, we can find that we can greatly reduce the personal storage pressure for RDF datasets by utilizing the cloud storage platform. The compression rate is not proportional to the number of triples. This is because the data content of each RDF datasets is different. On the whole, our experiment has achieved satisfactory results.

5. Conclusion

Our proposed solution not only achieves high performance lossless compression but also provides an open cloud storage platform. Due to our processing is done over the network, the user experience is not as good as centralized storage approach. But we believe that with the development of the network this gap will become more and more small. In terms of the storage security, our cloud storage approach is more secure compared to traditional compression approaches. Because subjects, predicates, and objects are stored in databases separately in the cloud, and their logical relationships are stored in the personal server, the eavesdropper cannot obtain complete information even if the cloud data is hacked. Security is one of the important factors for developing Semantic Web. Especially, the cloud platform provides standard interface to the update and delete data, it means that we don't need to access the personal server frequently. It is useful for IoT(Internet of Thing) devices that have less processing capacity and data is dynamic.

References

[1] Bizer C, Heath T, Idehen K and Berner-Lee T 2008 *Linked Data on the Web (LDOW2008)* Beijing China p21
[2] LOD (linked open data) sets. From website: http://lod-cloud.net/.
[3] Javier D. F, Claudio G and Miguel A. M 2010 *RDF Compression: Basic Approaches* p 1091
[4] Fern´andez J D, Mart´ınez-Prieto M A, Guti´errez C, Polleres A and Arias M 2013 *Binary RDF representation for publication and exchange (HDT). Web Semantics: Science, Services and Agents on the World Wide Web* p22
[5] Joshi A, Hitzler P and Dong G 2013 *Logical linked data compression. In The Semantic Web: Semantics and Big Data* p 170
[6] Fielding R T, 2000 *Representational State Transfer (REST): Architectural Styles and the Design of Network-based Software Architectures*. University of California, Irvine.
[7] Jens L, Robert I, Max J, Anja J, et al. 2012 *DBpedia - A Large-scale, Multilingual Knowledge Base Extracted from Wikipedia* pp 1-5
[8] Urbani J, Maassen J, Drost N, Seinstra F and Bal H 2013 *Scalable RDF data compression with MapReduce* P 24
[9] Dean J and Ghemawat S M 2004 *simplified data processing on large clusters* In Proceedings of the USENIX Symposium on Operating Systems Design & Implementation (OSDI) p 137–147

Acknowledgments
This research was supported by Basic Science Research Program through the National Research Foundation of Korea (NRF) funded by the Ministry of Education (No. 2016R1D1B02008553).

Design of an API recommendation system in android programming

Jinyang Liu[1]* and Zhiyi Ma[2,3,4]

[1]Academy for Advanced Interdisciplinary Studies, Peking University, Beijing, China
[2]School of Electronics Engineering & Computer Science, Peking University, Beijing, China
[3]Key laboratory of High Confidence Software Technologies (Peking University), Ministry of Education, Beijing, China
[4]Author to whom any correspondence should be addressed mazhiyi@pku.edu.cn
*jyliu.aais@pku.edu.cn

Abstract. Android has been the most popular operation system in the smartphone field, and Android programming is an import branch in programming work. When dealing with programming tasks, there are many APIs that are often used by programmers, and it costs much time for them to choose and check the usages of APIs repeatedly. This paper raises a design of API recommendation system in android programming, which uses AST tools in Java to extract key features of codes, and uses tf-idf, LDA and API2Vec models on the data to generate recommendation results. +The experiment results—— at maximum 80%+overall recall rate and nearly 50% top-10 recall rate prove the efficiency of the system.

1. Introduction

Over the past decades, smart phones have been a most popular goods all around the world, and the systems and applications running on the phone platforms also have become a research hotspot of computer science. As the main, or in a sense, "the only" open source operation system of smart phone platform, Android system has developed a lot during these years. According to statistic data from Kantar[1], 77.4% smart phones in China and 59.1% of them in US is in Android system at March 2018. As a huge amount of people are working on Android, to design and develop new applications, there is a strong need of them which is not resolved, that is, a systematic and effective algorithm or tool which can help them in coding. As we all know, to write a program file even just a code sequence is a hard mission that needs knowledge of programming language, the comprehension of functions, and logical reasoning. So, nowadays computer cannot fully complete a programming task without any human's participation (expect for some easy and patterning tasks). However, many researches have been made on the topic of computer doing programming work as an assistant of human. In many popular IDEs (Integrated Development Environment), the software could point out grammar mistakes of the codes, and list out the existing variable and function names to save programmers' time for typing them again and again. Early stage of the research focuses on grammar and structural attributes of the code, and when the generation of data arrived, recent researchers do more work and make more progresses on the content and semantics of codes, to let computers comprehend more about the function and objective of them. With trained model on big data, new algorithms may have the ability of complete code sequences based on context, and recommend coding elements for programmer in the deduction of the algorithm the programmers may use after the existing code fragment.

This paper provides a model for Android-programming that can generate recommendations for programmers of API usage, which is a content-based method and designers have tested several data-processing models on it. The model contains three parts: one is a module of transferring codes into a new data form for computer to process with, the second is a model to learn and develop with the data, and the last is a tool to generate coding sequences or recommendations based on the results of the model. The rest part of this paper is organized as follows: Part 2 introduces existing works of others in the area, part 3 describes the main parts of the model—— preprocessing for code files and learning model of processed data. Part 4 is a display and analysis of the experimental result, and part 5 is a conclusion.

2. Related Work

Many researches have been made in recent years, and this paper could only introduce some representative ones. Back to 2000, Ye et al. [2][3] designed a programming assisting system called CodeBreaker, which can recommend code sequences for users. It is based on a searching technology, and Bajracharya et al.[4] at 2010 did improved work on the same topic. In 2017, Nyugen[5] use Word2Vec on APIs in the codes to made a new vector representation of codes in searching or other methods of recommendation. In 2015, Lu et al.[6] raised a new searching technics based on WordNet, which has a good experiment result. Focusing on API use recommendation, Stylos [7] designed a tool of searching API class, methods and use examples called Mica. Chan et al.[8] give an API searching method based on a graph model representing the relationship between APIs and their invocations. On the strategy of API recommending, Bruch et al.[9] raised 3 common strategies, and Proksch et al.[10] designed a model of Bayesian Network. Nguyen[11] developed a tool of API recommendation based on Hidden Markov Model(HMM), which is called DroidAssist. Also, he gave an API embedding method called API2Vec, and this paper will discuss about it later.

3. The recommendation model

To help readers understand the work and model of this paper better, figure 1 is a complete sketch of our system. Firstly, the paper will introduce the code data preprocessing unit in our model.

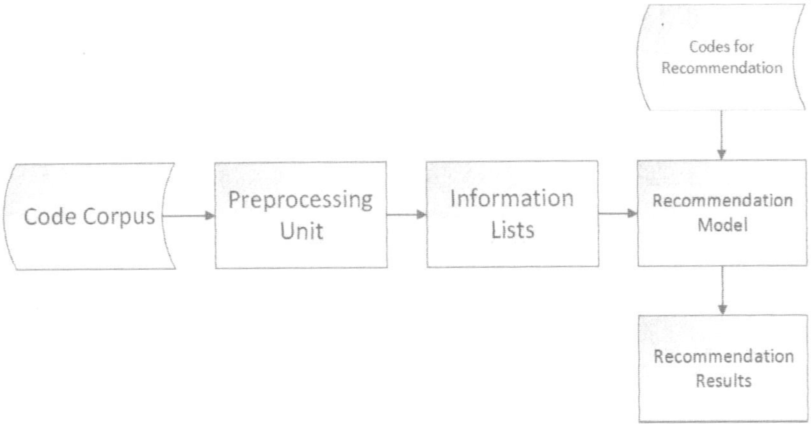

Figure 1. The whole structure of our recommendation system.

3.1. Data preprocessing unit

Android programs are written in Java Language, and figure 2 shows a typical Android code fragment. It is easy to find out that a code contains plenty of information and features, and when code data is used in code recommendation system, designers need to extract some key attributes of code itself, as there are redundant information in the code that is irrelevant with the content and subject of the code, for example, names of local variables, structural statements (such as loop and determine statements),

IOP Conf. Series: Journal of Physics: Conf. Series **1195** (2019) 012003 doi:10.1088/1742-6596/1195/1/012003

and so on. In the model this paper provides, it extracts 2 kinds of information in the code: The imported packages and the invocation methods. This two kinds of information contains main semantic features of a code. To generate a new object of an existed class, can be regarded as to invoke a constructor of a class, so it is also in the information field the model takes account into. Other elements in the code, such as a definition of a local class, will be only used in at most one program set, so this model does not extract this kind of information. Figure 3 and table 1 shows the original Java code, and the corresponding data extracted from it in this method. After the extraction step, a code is transferred into a list of "imports" and "invocations" for models to process with, which can be called an *I-I list*. The elements of the list is sorted by the visit order of the grammar tree of the code, so it also includes some of the information of the order in which the element appears in the code (though a code is a tree structure other than a 1-dimension list, so the list could not describe the order of execution completely).

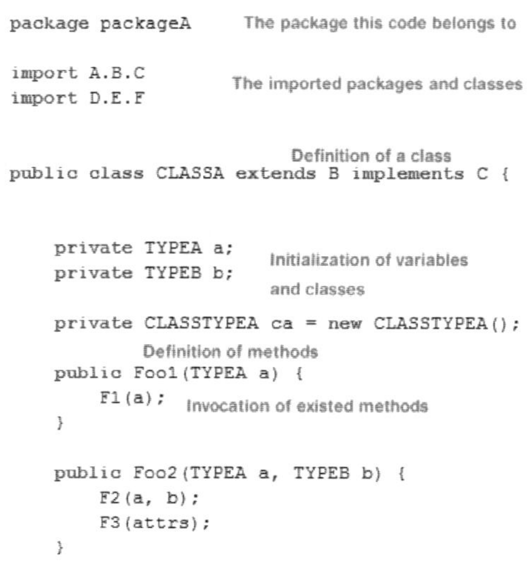

Figure 2. part of an Android code file and the instruction of some import elements.

Next, this paper introduces the method the model used in extraction step in detail. To extract the information needed, the model fulfills the task by generating, visiting and parsing the AST (Abstract Syntax Tree). The data structure AST represents codes in a tree form, and each node contains an element in the code, which is distributed in logical order and easy to be parsed. In Java Language, Eclipse provides useful classes in JDT for programmer to view and extract the information of grammar/syntax tree in the org.eclipse.jdt package[12], and with the ASTNode classes and the ASTVisitor Class, a parser tool is designed that can parse the syntax tree of any Java codes, and generate the lists of imports and invocations from them. With looking into the AST information deeply, our tools can distinguish the methods that have the same name but come from different fields. The classes and methods that is defined in a single android project is abandoned from our data extraction, for example, an imported class whose name is started with 'com.' . Figure 4 is an example of extraction.

```java
package com.ragnarok.javaparser;

import com.sun.istack.internal.Nullable;
import java.lang.Override;

public class User {

 @Nullable
 private String foo = "123123";
 private Foo a;

 public void UserMethod() {}

 static class Foo {
  private String fooString = "123123";

  public Foo FooMethod() {
    Foo g;
    return g;
  }
 }
 public void TestMethod(){
    Foo f;
    f.FooMethod();
}
}
```

Figure 3. The original code.

Table 1. Code elements extracted from the code in figure 3. The red lines are the elements used in building the I-I list.

Type	Name	Field
Import	com.sun.istack.internal.Nullable	
Import	java.lang.Override	
Class	User	
Field	String	foo
Field	Foo	a
MethodDeclaration	UserMethod	User
Class	Foo	
Field	String	fooString
MethodDeclaration	FooMethod	Foo
MethodDeclaration	TestMethod	User
MethodInvocation	FooMethod	Foo

```java
import org.eclipse.jdt.core.dom.CompilationUnit;

import com.ast.util.JdtAstUtil;
import com.ast.visitor.DemoVisitor;

import java.io.FileNotFoundException;
import java.io.FileOutputStream;
import java.io.PrintStream;
public class DemoVisitorTest {

    public DemoVisitorTest(String path,String destpath)
            throws FileNotFoundException {
        CompilationUnit comp = JdtAstUtil.getCompilationUnit(path);

        DemoVisitor visitor = new DemoVisitor();
        PrintStream ps=new PrintStream(new FileOutputStream(destpath));
        System.setOut(ps);
        comp.accept(visitor);
    }
    public static void main(String[] args)
            throws FileNotFoundException {
        String file=args[0];
        String destfile=args[1];
        new DemoVisitorTest(file,destfile);
    }
}
```

Import: org.eclipse.jdt.core.dom.CompilationUnit
Import: java.io.FileNotFoundException
Import: java.io.FileOutputStream
Import: java.io.PrintStream
Invocation: JdtAstUtil.getCompilationUnit
Invocation: PrintStream.new
Invocation: FileOutputStream.new
Invocation: System.setOut
Invocation: CompilationUnit.accept

Figure 4. An example of the extraction process, the code is on the left and the extracted features are on the right. Reader can find that some features of this code have been abandoned.

After that, the model needs to do filtering work on the lists: just like many NLP (Natural Language Processing) tasks, elements that appear below certain times would be erased from data. To make our experiment more convincing, several thresholds of appearing times are chosen to compare the results.

The database this paper used is collected from GitHub, partly from the GitHub Java Corpus dataset and partly is collected by the designers. It has 95,505 code files in all. After the extraction, some statistics data is listed in figure 5. From the statistics data, it can be found out that there are a huge number of APIs and their usages/invocations in the dataset, and the times of each element is used are

subjected to a typical long-tail distribution. Comparing with the average length of the I-I lists of codes, the data to be used is really sparse, which is a great challenge.

3.2. Data processing and training models

Part 3 talks about the preprocessing step in figure 1, and in this part this paper will introduce the key part in our model: the model used to learn on the data fetched from codes. As is mentioned before, the goal of this paper is to build an API recommending method based on the content other than grammar rules. Thus, 3 semantic models in NLP are tested on the model to get a vectorized representation of the codes for recommendation. In the recommending step, a code is vectorized with the model, the most similar codes in database are found as the recommendation. The models this paper use are tf-idf model, LDA model and API2Vec model. These models provide varied vector representations of codes, which are frequently used in similar tasks.

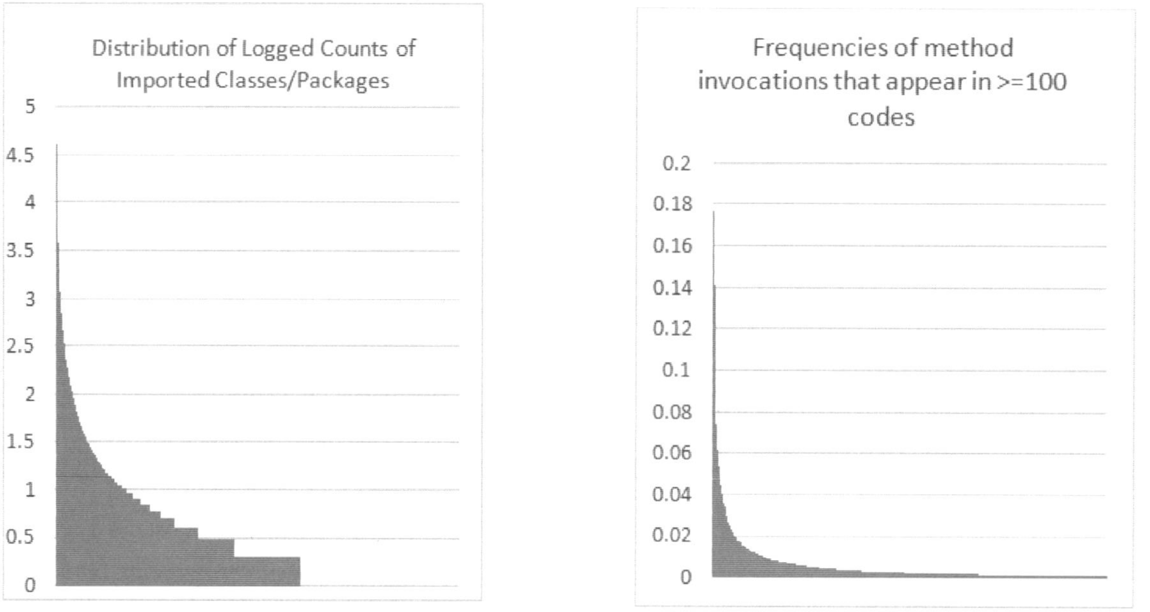

(a) (b)

Figure 5. (a)Distribution of Logged Counts of Imported Classes/Packages. Y axis is the log10 value of appearing counts. About 40% imported terms only appear in one code file. (b) Frequencies of method invocations that appear in >=100 codes. The distribution is much sparser than import terms.

3.2.1. tf-idf Model. As one item may appear in an I-I list for many times, to use a normal one-hot vector representation for an I-I list is not adequate for our task. To take the appearing times and total appear frequency of one term into account, tf-idf model can be used to make a vector representation of an I-I list. If one uses the traditional aspect of recommending system, to regard one code as a user and one term as an item, the tf-idf vector can give the “interest” of one code on its terms, with the weights of each term as its components.

The weight of each term in a document (in the task, the “document” means the I-I list) in tf-idf model is calculated as following [13]:

$$tfidf(t, d, D) = tf(t, d) * idf(t, d) \tag{1}$$

where *tf* is called term frequency and *idf* is inverse document frequency. *tfidf(t,d,D)* is the tf-idf weight of term t in document d out from corpus D.

tf is the frequency of one term in the document:

$$\mathrm{tf}(t,d) = \frac{f(t,d)}{\sum_{t'\in D} f(t',d)} \tag{2}$$

f(t,d) is the count/frequency that term t appears in document d.
idf is a modifying factor that evaluating the popularity of one term in all the documents:

$$\mathrm{idf}\ (t,d) = \log \frac{N}{|\{d \in D : t \in d\}|} \tag{3}$$

N is the total amount of documents, and the denominator of log is the count of documents that t appears in it. The more popular a term is, the smaller corresponding idf factor is. idf factor is used to balance the weights of terms between popular ones and unpopular ones. For example, if A is a very popular term that appears in many codes, B is a term that is rarely used, and they both appear in a code at the same time, the "interest" of this code on B should be much more than A, which would be reflected on the tf-idf weights. This feature is very important for the task, as there are really popular package and methods in our database, and much more rarely-used terms.

After tf-idf vectorization, an I-I list is transferred into an N-dimension vector, where N is the total amount of terms, and each component is the weight/interest of one term in the code. The model will use these vectors to rank the recommendation results of single terms, and can calculate the similarities between the codes, where other methods will also be adopted.

3.2.2. LDA. To calculate the similarities between codes, except for directly using terms in the I-I lists, the model can also extract some semantic features from the code, for example, the "topics". There have been many topic models, such as LSA (Latent Semantic Analysis), p-LSA (probabilistic Latent Semantic Analysis), and LDA (Latent Dirichlet Allocation). The last one is the topic model the model like to adopt. LDA is raised by David Blei et al. [14], which is in some extent an improvement of p-LSA model. It regards documents as mixed-model of topics, and each word is corresponding to a distribution vector of topics, that is, each word has a contribution factor to each topic. Actually, LDA is a directed graph model with the theory of Bayesian Estimation. After training the LDA model on the corpus, the model could generate a topic distribution for each code, and use the topic vector (the dimension is number of topics) to calculate similarities.

3.2.3. API2Vec. API2Vec, as the implication of its name, is a method derived from Word2Vec, the widely-used method in the vectorization of natural language data/vector embedding or words. Word2Vec[15] uses a neural network model called CBOW model to train a vector representation for words.. Nguyen et al. in [16] applied the Word2Vec on Java codes data, then examined and proved the availability and practicability of API2Vec that use API sets as input data. Figure 6 shows the CBOW model in Word2Vec and the application of this model into API learning. In the task from this paper, API2Vec model cam also be used to learn a vector for each term, then a matrix representation of code files can be formed for similarity calculation. For each code file, the normalized average vector of terms in it is used for vector representation of the whole file.

3.3. Generate recommendation results

The model of this paper used the following method for API-use recommendation: For an incomplete code, it uses one of the former methods to calculate the similarities between it and the codes already in the data corpus. The model chooses the K most similar codes of the incomplete one, and provides the APIs in them as recommendations. The APIs is ranked by the accumulated tf-idf weights in the K codes. In the next part, this paper will show the detailed experiment results in the database.

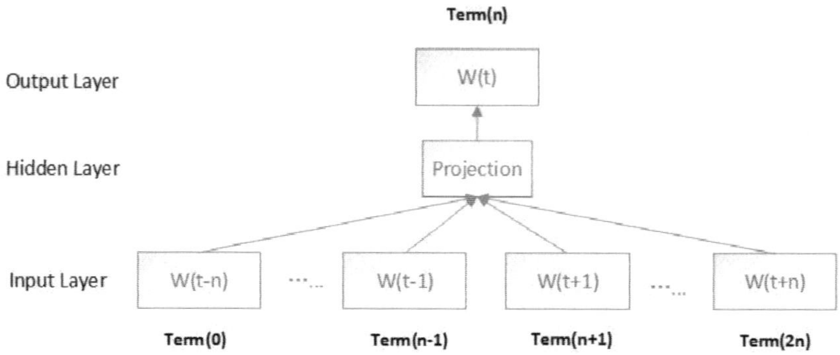

(a)

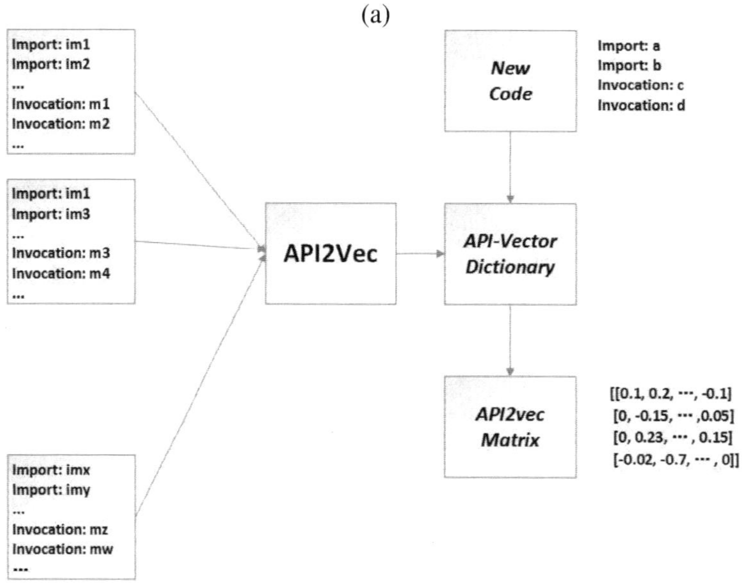

(b)

Figure 6. (a) CBOW Model and its application on API data. This figure shows a CBOW model with window length (2n+1).(b) How API2Vec works. The model is trained on code corpus, and generate matrix representation on any given code.

4. Experimental Results

The database this paper use has 95,505 code files, and the paper randomly select about 1,000 from them as test data, and train our models on the rest. When testing the model, on each test file, one term is randomly removed from its I-I list, and system will check if the recommendation result contains the lost one. The overall recall rates are listed in Table 2 with varied thresholds of how to remove barely-used terms. This paper use 10 as the value of factor K. For LDA and API2Vec models, the paper tested several parameter groups, and chose out the best ones to provide results. As for LDA, number of topics is chosen in 30, 50, 70 and 100. For API2Vec, the dimension of vectors is chosen from 100, 200 and 300, and the length of window is chosen between 5 and 10. The parameters are elected for the range in which earlier primal experiment show good performance.

Tabel 2. Overall recall rates of 3 methods.

	Overall Recall Rate(%)		
Threshold	tf-idf	LDA	API2Vec
3	79.84	72.94	**81.03**

5	81.05	71.19	**81.86**
10	80.23	73.12	**82.23**
20	79.92	73.02	**83.01**
50	78.05	71.77	**83.07**
100	79.02	72.72	**84.21**

As table 2 shows, tf-idf and API2Vec methods outperform the LDA model, and these two have close evaluation scores. To evaluate and compare tf-idf and API2Vec deeply, table 3 provides the top-N recall rates, that is, only return N terms that have most accumulated tf-idf weights in recommendation results.

Table 3. Top-N recall rates

	TopN Recall Rate(%)							
	Top5		Top10		Top20		Top30	
Threshold	tf-idf	API2Vec	tf-idf	API2Vec	tf-idf	API2Vec	tf-idf	API2Vec
3	**36.01**	30.1	41.77	**42.58**	49.07	**55.36**	54.63	**65.67**
5	**35.74**	30.41	42.12	**42.58**	48.3	**55.77**	53.24	**63.09**
10	**36.25**	32.64	43.36	**44.83**	50.77	**57.13**	56.23	**63.12**
20	**35.22**	31.92	41.5	**46.32**	47.89	**59.59**	53.35	**66.22**
50	**34.47**	33.02	41.82	**46.92**	49.38	**59.35**	54.24	**66.04**
100	**34.79**	34.53	43.2	**48.74**	50.47	**62**	55.45	**67.89**

On top-5 recall rates, API2Vec does not perform so well, but it has obviously better performance on other top-N recommendation results. In practical scene, a weakness of API2Vec is that it costs more time than tf-idf to calculate code similarities. Whatever, these two methods have proved themselves to be efficient in API recommendation task of Android programming.

From the experimental results, it is concluded that using API2Vec representation for code files in similarity calculation, and use tf-idf weights in term ranking performs best. The shortage of LDA model is that, when "topics" are extracted from data, the representation loses some features, where the topics may not describe the code fully. Also, the lengths of I-I lists in our task are varied, some of which are great, and the most of which are small. On short text data, conventional LDA model cannot perform as well as that on long text data.

As the thresholds of API used times increases, the precision of recommendation also increases naturally. Thus, if users focus on APIs in a certain domain, they can have good feedback from the system. With the expansion of database and optimization of parameters, API2Vec will show significant growth of effects. Nevertheless, due to huge amount of APIs in data, and the sparsity of distribution of them, there would be many redundant APIs on the recommendation rank list, which lowers the top-N recall rate. In future work, designers will deal with the problem by applying other technics, e.g., using grammar rules to erase some candidates, to improve the top-N recall rates.

5. Conclusion

This paper discusses about the API recommendation task in the Android programming. The paper builds a recommendation system, and evaluate several model units on it. A mixed model of API2Vec and tf-idf has a relatively good performance, but still has considerable improvement space in top-N recommendation recall rate. As mentioned before, our experiment and the real application scene both must face with a data source that is much sparser than the data in stereotype recommendation systems, which contributes to the great difficulty in this task. As for recommendation algorithms, some researchers have tried to use deep learning models independently or mix the network into Collaborative Filtering. However, deep learning model cannot deal with our task well as the number of terms need a network model with great number of parameters to be built, which challenges the

computing environment strongly, and questions the availability of itself. Also, deep learning methods, such as LSTM and AutoEncoder have not shown its considerable advantage over traditional algorithms. With the improvement of database size and computing environment, the result of the model and system in this paper will show more competitive power.

References

[1] Kantar Worldpanel, https://www.kantarworldpanel.com/
[2] Ye, Y., & Fischer, G. (2002). Supporting reuse by delivering task-relevant and personalized information. *International Conference on Software Engineering* (pp.513-523). IEEE.
[3] Ye, Y., Reeves, B., & Reeves, B. (2000). Integrating active information delivery and reuse repository systems. *Century Applications* (Vol.25, pp.60-68). ACM.
[4] Bajracharya, S. K., Ossher, J., & Lopes, C. V. (2010). Leveraging usage similarity for effective retrieval of examples in code repositories.
[5] Nguyen, T. N., Nguyen, T. N., Nguyen, T. N., Nguyen, T. N., & Nguyen, T. N. (2017). Combining Word2Vec with revised vector space model for better code retrieval. *Ieee/acm, International Conference on Software Engineering Companion* (pp.183-185). IEEE.
[6] Lu, M., Sun, X., Wang, S., Lo, D., & Duan, Y. (2015). Query expansion via WordNet for effective code search. *IEEE, International Conference on Software Analysis, Evolution and Reengineering* (pp.545-549). IEEE.
[7] Stylos, J., & Myers, B. A. (2006). Mica: A Web-Search Tool for Finding API Components and Examples. *Visual Languages and Human-Centric Computing* (pp.195-202). IEEE Computer Society.
[8] Chan, W. K., Cheng, H., & Lo, D. (2012). Searching connected API subgraph via text phrases. *FSE* (pp.1-11).
[9] Bruch, M., Monperrus, M., & Mezini, M. (2009). Learning from examples to improve code completion systems. *The, Joint Meeting of the European Software Engineering Conference and the ACM Sigsoft Symposium on the Foundations of Software Engineering* (pp.213-222). ACM.
[10] Proksch, S., Lerch, J., & Mezini, M. (2015). *Intelligent Code Completion with Bayesian Networks*. ACM.
[11] Nguyen, T. T., Pham, H. V., Vu, P. M., & Nguyen, T. T. (2016). Recommending API Usages for Mobile Apps with Hidden Markov Model. *Ieee/acm International Conference on Automated Software Engineering*(pp.795-800). IEEE.
[12] Eclipse documentation - Archived Release, http://help.eclipse.org/kepler/
[13] tf-idf, Wikipedia, https://en.wikipedia.org/wiki/Tf%E2%80%93idf
[14] Blei, D. M., Ng, A. Y., & Jordan, M. I. (2003). Latent dirichlet allocation. *J Machine Learning Research Archive, 3*, 993-1022.
[15] Mikolov, T., Chen, K., Corrado, G., & Dean, J. (2013). Efficient estimation of word representations in vector space. *Computer Science*.
[16] Nguyen, T. D., Nguyen, A. T., Phan, H. D., & Nguyen, T. N. (2017). Exploring API Embedding for API Usages and Applications. *International Conference on Software Engineering* (pp.438-449). IEEE Press.

Acknowledgements

This work is supported by the National Natural Science Foundation of China (No. 61672046).

Test case generation for WS-BPEL from a static call graph

Wareerat Bousanoh and Taratip Suwannasart

Department of Computer Engineering, Faculty of Engineering
Chulalongkorn University, Bangkok, Thailand

Email: Wareerat.Bo@student.chula.ac.th, Taratip.S@chula.ac.th

Abstract. Web service is reusable, easy to maintain, widely used, and very popular. However, web service development is not adhered to any development languages in variety of platforms or frameworks. There must be a standard in controlling and coordinating a business process through a web service. This standard is called WS-BPEL. Web service testing is an important process in software development to meet the needs of business process. Recently some researches proposed concepts and tools used to generate test case for WS-BPEL, while they focus only a business processes within a single WS-BPEL file. This paper proposes an approach for generating a WS-BPEL test cases from a static call graph to support the case which has a calling subprocess between WS-BPEL files. The generated test cases are complied with test paths based on the branch coverage.

1. Introduction

Nowadays, web service is very popular, due to its reusable and easy to maintain. A web service is designed to support interoperable between software applications over a network. Web service development is one of the software development under the principle of Service Oriented Architecture (SOA) [1]. SOA is about dividing and making software components independence. However, web service development is not adhered to any development languages. For that reason, there must be a web standard in controlling and coordinating a business process through a web service. The standard is called Web Services Business Process Execution Language (WS-BPEL), which is developed by OASIS organization [2].

WS-BPEL is a language for describing a behaviour of a business process based on interactions between the process and its partners. The interaction with each partner occurs through web service interfaces [2]. A WS-BPEL process represents all partners and interactions with these partners in terms of abstract WSDL interfaces [2]. WSDL is used to describe web services and tells a client how to compose a web service request and describes the interface that is provided by the web service provider.

Web service testing is an important process in software development to meet the needs of business process. It also reduces the number of software errors before delivery and ensures that a software is correct. Recently, there is a research which proposes concepts and a tool to generate test cases for WS-BPEL, while it focuses only a business processes within a single WS-BPEL file. In case of calling a subprocess between WS-BPEL files, a static call graph is used to help testers to understand WS-BPEL files and related variables in test paths for generating test cases. Otherwise, testers must be proficient in WS-BPEL files for generating test cases.

This paper proposes an approach for generating WS-BPEL test cases from a static call graph to support the case which has a calling a subprocess between WS-BPEL files. The generated test cases are complied with test paths based on the branch coverage.

2. Related work

There have been researches attempting to generate test cases for WS-BPEL. P. Nakngern and T. Suwannasart [3] proposed a design of WS-BPEL test case generation tool based on path conditions, which selects test paths from the control flow graph and map all related variables with random values from path conditions and their constraints. Ebrahim Shamsoddin-Motlagh [4] presented an approach for test cases generation automatically at the SOA system, which creates a control flow graph of BPEL file in the system and services related to the main service. Then, the test cases are created randomly for primary test from the graph in the generated system. S. Laokok and T. Suwannasart [5] proposed an approach to generate test cases from a static call graph in order to cover all class interfaces based on branch coverage.

3. Background

3.1. WS-BPEL

Web Service Business Process Execution Language (WS-BPEL) is the language for determining a model describing behavior of business processes which are coordinated between users and related web services [2]. It is a standard language developed by OASIS, and its distribution started in 2007 in a form of XML structure.

A business process of WS-BPEL is a reusable by using subprocess extension [6]. A subprocess is a part of BPEL code that is separated from a process which has the same behavior. The subprocess extension has the following benefits:

- to reduce a duplicate process.
- easy to modify and maintain.
- to reduce memory usage, especially for a complex process.

In this paper, we support for two types of subprocesses, which are inline subprocess and standalone subprocess.

1) An inline subprocess is defined as a part of BPEL 2.0 process at the <process> level [6].
2) A standalone subprocess is the process defined in a file with extension .sbpel (subprocess BPEL extension) [6].

3.2. Static call graph

A static call graph is a graph which represents the relation between methods [7]. Each method in a program is a node where a relationship between nodes is an edge. An activity in BPEL files is similar to a method in the general program, thus each node represents an activity in BPEL files and an edge represents calling of a BPEL file. For example, figure 1 shows a call graph for a simple Java program that has method A calling method B, method B calls method C, and method C calls methods B and D.

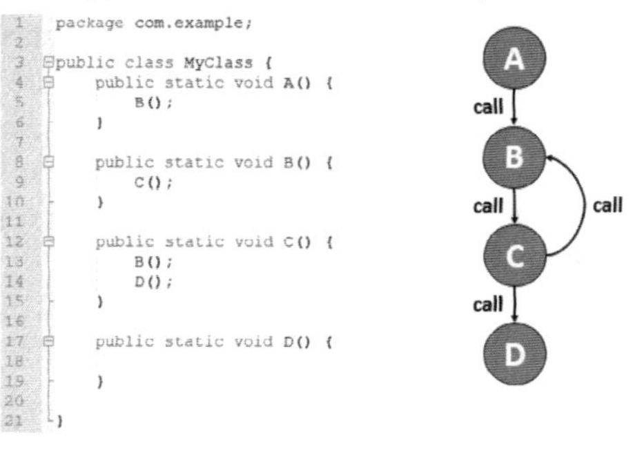

Figure 1. An example of a static call graph in Java program [7].

IOP Conf. Series: Journal of Physics: Conf. Series **1195** (2019) 012004 doi:10.1088/1742-6596/1195/1/012004

4. Methodology

This paper presents an approach for generating WS-BPEL test cases using a static call graph. The framework of our approach is shown in figure 2 and is described in subsections as follows:

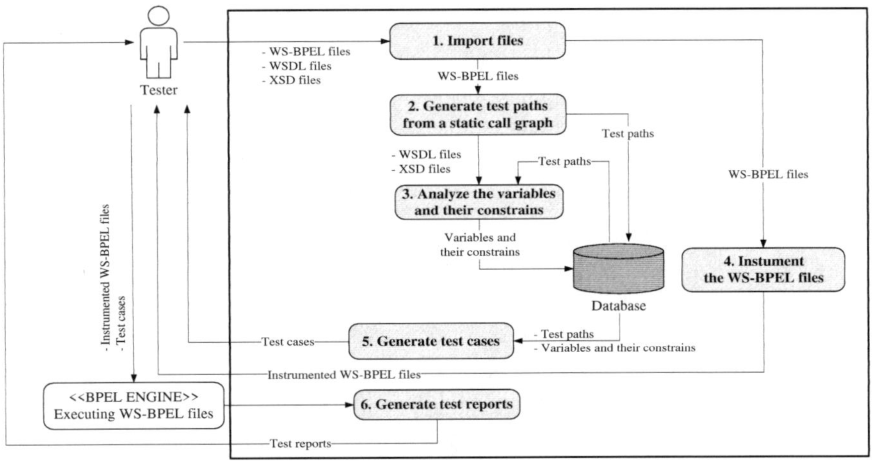

Figure 2. An overview of test case generation method.

4.1. Import files

Testers import files which consist of WS-BPEL, WSDL, and XSD files.

4.2. Generate test paths from a static call graph

This process has four steps. Firstly, all XML tags in WS-BPEL files are read to generate a static call graph in the next step. The WS-BPEL example file, CreditCardPay.bpel, a promotion of credit card payment in a shopping mall, which is the initial WS-BPEL file. The XML tag <bpelx> in CreditCardPay.bpel file calling the sbp_VISARewards.bpel file is shown in figure 3.

```
<assign name="assign_VISADiscount">
  <copy>
    <from>$inputVariable.payload/client:Price-((5*$inputVariable.payload/client:Price)/100)</from>
    <to expressionLanguage="urn:oasis:names:tc:wsbpel:2.0:sublang:xpath1.0">$outputVariable.payload/client:NetPrice</to>
  </copy>
</assign>
<extensionActivity>
  <bpelx:call name="sbpVISARewards" xmlns:sp1="http://xmlns.oracle.com/SOACSdemo/CreditRewards/sbp_VISARewards"
              target="sp1:sbp_VISARewards"/>
</extensionActivity>
```

Figure 3. The Initial WS-BPEL file. (CreditCardPay.bpel)

The XML tag <bplex> in sbp_VISARewards.bpel file which calls the sbp_VISAGift042018.sbpel file is shown in figure 4, and the sbp_VISAGift042018.sbpel file is shown in figure 5.

```
<assign name="assCashBack1000">
  <copy>
    <from>1500</from>
    <to expressionLanguage="urn:oasis:names:tc:wsbpel:2.0:sublang:xpath1.0">$CashbackAmount_Output</to>
  </copy>
</assign>
<extensionActivity>
  <bpelx:call name="sbpVISAGift042018" xmlns:sp1="http://xmlns.oracle.com/SOACSdemo/CreditRewards/sbp_VISAGi
              target="sp1:sbp_VISAGift042018"/>
</extensionActivity>
```

Figure 4. The sbp_VISARewards.sbpel file.

```
<sequence name="main">
  <receive name="receiveInput" partnerLink="serviceVISAGift042018"/>
  <if name="price">
    <documentation>
      <![CDATA[25,000>price>=10,000]]>
    </documentation>
    <else>
      <documentation>
        <![CDATA[price>=25,000]]>
      </documentation>
      <sequence name="Sequence1">
        <assign name="assGiftVoucher"></assign>
        <invoke name="Invoke1" bpelx:invokeAsDetail="no" partnerLink="serviceGiftVoucher"/>
      </sequence>
    </else>
  </if>
</sequence>
```

Figure 5. The sbp_VISAGift201804.sbpel file.

Secondly, each WS-BPEL activity from CreditCardPay.bpel, sbp_VISARewards.sbpel, and sbp_VISAGift201804.sbpel file are analyzed to generate a control flow graph. For Example, the control flow graph of CreditCardPay.bpel file, sbp_VISARewards.sbpel file and sbp_VISAGift201804.sbpel file are shown in figure 6, 7, and 8 respectively.

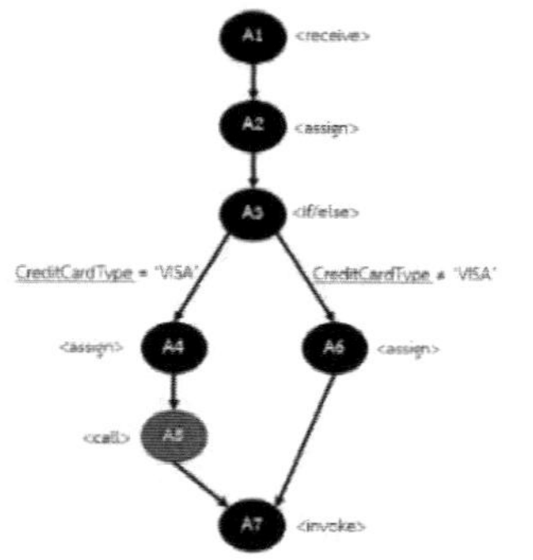

Figure 6. The control flow graph of CreditCardPay.bpel file.

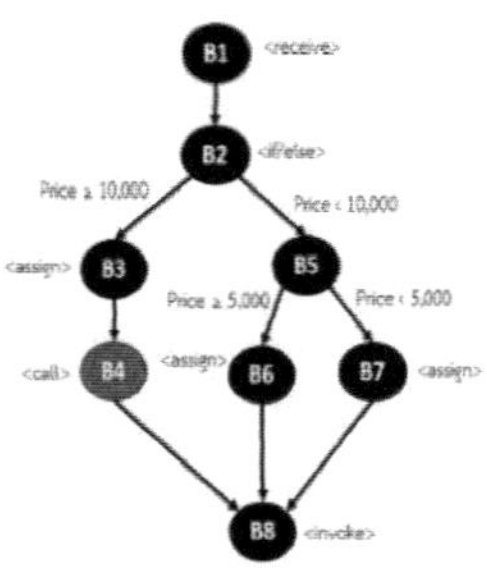

Figure 7. The control flow graph of sbp_VISARewards.sbpel file.

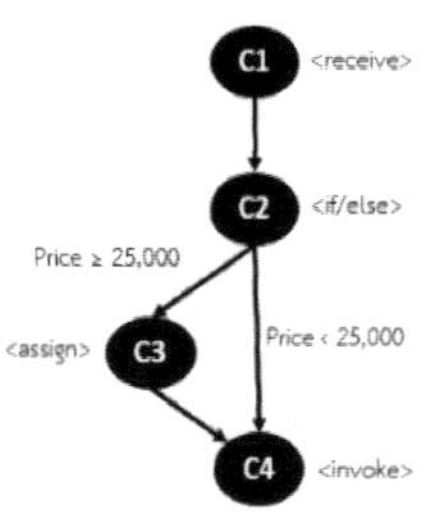

Figure 8. The control flow graph of sbp_VISARewards.sbpel file.

Thirdly, a static call graph is generated from each control flow graph. The example in figure 9 presents control flow graphs of each activity and created static call graph of business process.

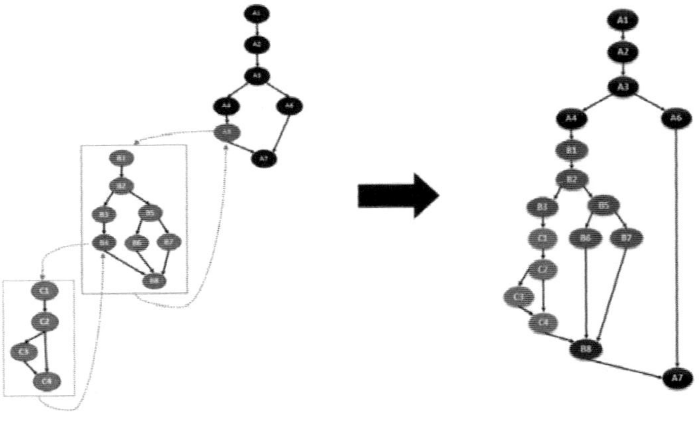

Figure 9. Control flow graphs of each activity and created static call graph of business process.

Finally, test paths are generated from the created static call graph and stored into the database. From the static call graph in figure 9, there have five test paths covering branch coverage as follows:

- P1: A1 – A2 – A3 – A4 – B1 – B2 – B3 – C1 – C2 – C3 – C4 – B8 – A7
- P2: A1 – A2 – A3 – A4 – B1 – B2 – B3 – B6 – C2 – C4 – B8 – A7
- P3: A1 – A2 – A3 – A4 – B1 – B2 – B5 – B6 – B8 – A7
- P4: A1 – A2 – A3 – A4 – B1 – B2 – B5 – B7 – B8 – A7
- P5: A1 – A2 – A3 – A6 – A7

4.3. Analyze the variables and their constrains

The variables and their constrains of each node of the static call graph are analyzed using the WSDL and XSD files which are imported from *process A* and the test paths generated from *process B*. Then, it is stored into the database.

4.4. Instrument the WS-BPEL files

In this process, the WS-BPEL files are instrumented to verify if all test cases cover all branches in the static call graph. At the end of this process, a tester will receive the instrumented WS-BPEL files to be executed through the BPEL engine.

4.5. Generate test cases

The test paths, variables and constrains which are outputs of *process B* and *C* and used to generate test cases based on branch coverage. The test input values are randomized by using the variable constrains. At the end of this process, a tester will receive the test cases to be executed through the BPEL engine.

4.6. Generate test reports

After the tester executes the instrumented WS-BPEL files and the test cases through the BPEL Engine, test logs are recorded by BPEL Engine. Finally, the test reports are generated using the test data from the previous processes.

5. Conclusion

This paper introduces a design of test case generation for WS-BPEL from a static call graph based on branch coverage. We propose an approach for generating test case that supports the case of calling subprocess between WS-BPEL files. In the future, we will aim to apply this approach in developing a tool from our proposed approach.

IOP Conf. Series: Journal of Physics: Conf. Series **1195** (2019) 012004 doi:10.1088/1742-6596/1195/1/012004

References

[1] "Reference Architecture Foundation for Service Oriented Architecture Version 1.0". [Online]. Available: http://docs.oasis-open.org/soa-rm/soa-ra/v1.0/cs01/soa-ra-v1.0-cs01.html [Accessed: 24 January 2018]

[2] "Web Services Business Process Execution Language Version 2.0". [Online]. Available: http://docs.oasis-open.org/wsbpel/2.0/OS/wsbpel-v2.0-OS.html [Accessed: 14 February 2018]

[3] P. Nakngern and T. Suwannasart (2017). A Design of WS-BPEL Test Case Generation Tool Based on Path Conditions. International MultiConference of Engineers and Computer Scientists.

[4] E. Shamsoddin-Motlagh (2013, October). Automatic Test Case Generation for Orchestration Languages at Service Oriented Architecture. International Journal of Computer Applications, Vol. 80, No7, pp. 35-44.

[5] S. Laokok and T. Suwannasart (2017). An Approach for Test Case Generation from a Static Call Graph for Object-Oriented Programming. International MultiConference of Engineers and Computer Scientists.

[6] "Oracle® Fusion Middleware Developer's Guide for Oracle SOA Suite". [Online]. Available: https://docs.oracle.com/middleware/1221/soasuite/develop/GUID-FAA53226-470D-4120-95ED-1EAB6D96AA3F.htm [Accessed: 14 February 2018]

[7] "Call Graph Construction Algorithms Explained". [Online]. Available: https://ben-holland.com/call-graph-construction-algorithms-explained/ [Accessed: 14 April 2018]

IOP Conf. Series: Journal of Physics: Conf. Series **1195** (2019) 012005 doi:10.1088/1742-6596/1195/1/012005

A data mining approach for creating a job position in the system for evaluating competencies

Ondrej Pektor[1], Bogdan Walek[1]* and Ivo Martinik[2]

[1] Department of Informatics and Computers, University of Ostrava, Ostrava, Czech Republic
[2] Department of Computer Science, VSB-Technical University of Ostrava, Ostrava, Czech Republic

*bogdan.walek@osu.cz

Abstract. This paper focuses on a data mining approach for automated retrieval of job position competencies based on a given job title and keywords that represent important competencies or concepts that are associated with a given position. The main aim is to retrieve and process the content of relevant job vacancies on the selected job portal. The output is a list of relevant words found along with their occurrence in crawled job vacancies that indicate important competencies that are required for a given job position. The HR manager then obtains a list of relevant competencies for the selected job position and the selected competencies can be added to the profile of job position in the system. Additionally, for the HR manager, the output is also a proposal to remove competencies from the job position profile, because they are not relevant (not found in searched job vacancies). These outputs are then a support tool for HR managers to create a list of all the competencies of a given job position and to store them in the job position profile. The implemented approach is verified on a specific example and the results are presented.

1. Introduction

Nowadays, the area of recruiting suitable and competent staff to job positions is a very current issue. For all companies, it is crucial to find workers who will not only be loyal and reliable but will, above all, have all the required competencies required for the job.

In companies, it is emphasised that job positions are occupied by competent employees. Competencies are defined as knowledge, skills and attributes that distinguish an expert from a layman. Another definition is that competency is a construction which helps define the level of skill and knowledge [1] [2]. A person competent to take up a particular job must meet the given competencies to a minimum. This information is called the competency model. It can be successfully used to define which employee skills are crucial for the job position [3] [4].

For instance, based on the evaluation of the job applicant competencies, the overall job position suitability can be determined.

Evaluating the employee's competencies can take place in various situations. For instance, in decision-making and recruitment of job applicants or regular evaluation of existing staff. The assessment of each competency can take the form of a language expression, years of work experience or percentage. In the case of the number of years of practice, this rating can be converted to percentages according to

the given reference value (maximum, minimum, etc.) and in the case of a language expression, the conversion to percent is easy if the number and order of language expressions are fixed.

This paper focuses on a data mining approach for automated retrieval of job position competencies based on a given job title and keywords that represent important competencies or concepts that are associated with a given position. The main aim is to retrieve and process the content of relevant job advertisements on the selected job portal. The output is a list of relevant words found along with their occurrence in crawled job advertisements that indicate important competencies that are required for a given job position. The HR manager then obtains a list of relevant competencies for the selected job position and the selected competencies can add to the profile of job position in the system. Additionally, for the HR manager, the output is also a proposal to remove competencies from the job position profile, because they are not relevant (were not found in searched job vacancies). These outputs are then a support tool for HR managers to create a list of all the competencies of a given job position and to store them in the job position profile.

2 Data mining approach

The structure of the proposed data mining approach is shown in the following figure:

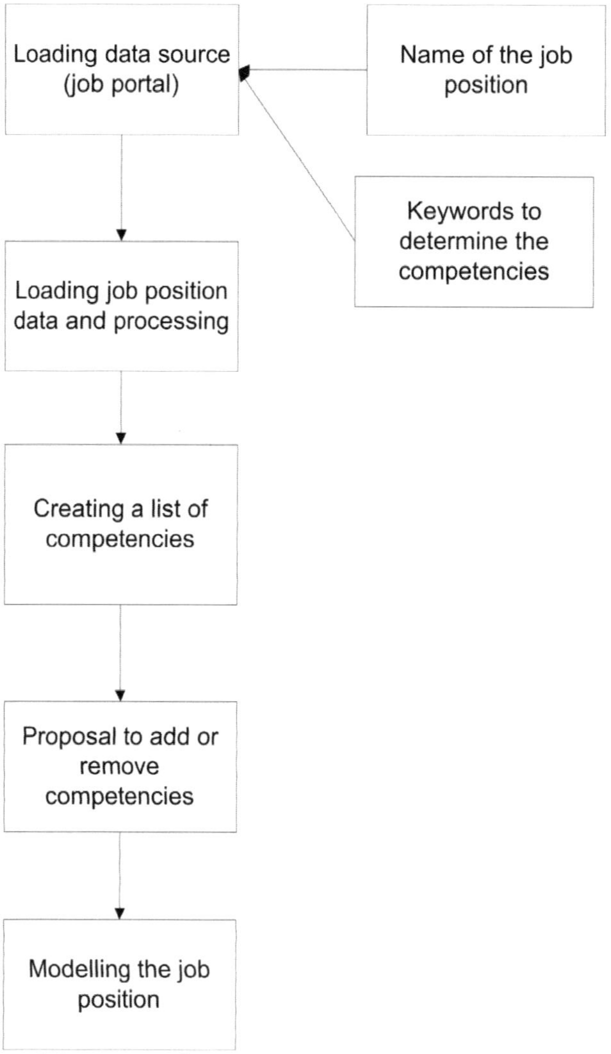

Figure 1. Proposed data mining approach.

Particular steps and functionalities of the proposed approach are described in more detail in the following subchapters.

2.1 Creating the name of the job position and keywords to determine the competencies
In the first step, the name of the job position is created. Creating the right name of the job position is very important because this name will be used to search job advertisements on the job portal. Job vacancies on the job portal will be searched based on the job title you have entered. In addition, it is important to include keywords in the system that indicate important competencies or concepts related to the given job position. An example may be an "Administrative assistant" job position. Keywords for this position can be: work with PC, Office, Word, Excel, etc.

2.2 Loading data source - job portal
In the next step, a data source is entered into the system by the administrator as a URL to the job portal. One or more data sources can be entered into the system, each represents the selected job portal.

2.3 Loading job position data and processing
Based on the job position title and the given data source, an automated search of all relevant advertisements is performed. Each advertisement is typically represented by an HTML page that contains its description under the advertisement name. The data mining module then looks for the relevant competencies in advertisement description. Each advertisement is then categorised and divided into words and phrases. Then the competencies are extracted and the list of relevant competencies is created.
The data mining algorithm is described below:

1. Creating a request to display relevant job vacancies on the job portal using the job title
2. Analysis of HTML response - gather links to individual job vacancies and find out the number of pages of all job vacancies
3. Sending requests for each HTML page and gather links to individual job vacancies
4. Submit a request for each of the collected links to job vacancies
    a. Analysis of the HTML code of the job vacancy - analysis of the job vacancy structure and detection if the job vacancy is in the Czech language - if not, the job vacancy is excluded
    b. Searching for the keywords entered in the job vacancy, if no keyword is found, the job vacancy is excluded
    c. The job vacancy description is divided into individual words
        i. Every word is processed and system detects if the word is on the list of excluded words and if it consists only of the letters, if not, the word is skipped
        ii. Identifying competencies that are already added to the work position and that contain the word being processed (these competencies will not be proposed to be deleted)
        iii. If statistics for that job vacancy do not include this word, the word is added if the word contains statistics, it increases the number of occurrences
        iv. Searching for a database of competencies that are not assigned to a job position and comparing it with a word being processed. If a match is found, the competency is added to the list of existing competencies recommended to add to the given job position (if the competency added is already added, its frequency counter will be increased)
        v. Comparing the word with the competency register from the database If a match is found, the word is added to the new competencies, if it has already been added, its occurrence counter

5. Suggested keywords are extracted from the statistics of all job vacancies if their occurrence is higher than 2

2.4 *Creating a list of competencies*

Based on the data from the previous step, the system shows the outputs directly to the user (typically HR manager). And the user can work immediately with the found competencies.
We distinguish two basic types of outputs:

- List of found competencies that occurred in the system database and in the found job vacancies, but are not part of the currently searched position in the system
- List of new proposed competencies - a list of all new competencies and keywords that a user can add to their own modelled job position along with the number of occurrences of a given competency

2.5 *Proposal to add new competencies and remove the competencies which are not found*

Based on the previous step, the user can decide which newly-proposed competencies will be inserted into the modelled position and removed from the modelled position. The advantage is that the data mining approach proposed by us is in the form of a module directly integrated into the system for evaluating competencies. The user can drag and drop the competencies into a modelled position and assign the importance for later evaluation. A comprehensive system for evaluating competencies that includes the data mining approach proposed in this article, is more described in these publications [5] [6].

2.6 *Modelling the job position*

In the final step, the user creates the final list of all relevant competencies. Then the user saves the modelled job position and the job position is ready for later evaluation.

3 Verification

For verification, a part of the web system has been created and contains all the necessary components. The web system is built on the .NET Core platform, so it is a multiplatform. As a data warehouse, the PostgreSQL database management system is used and the Object-Relational Entity Framework is used to access it. The Frontend system runs on ASP.Net Core MVC using responsive framework Bootstrap.

The fuzzystring library, authored by Kevin D. Jones, was used to compare the words of the ad and the Competence Register. [7].

To validate the proposed approach, a job was created in the system called Programmer. For this position, the following competencies were added to the system:

- Analytical reasoning
- Technical ability
- Working with computer
- Communicativeness
- Compliance with corporate culture

Then, there is competency dictionary created in the system, which contains a list of pre-created competencies that can be used for insertion into different job positions. The HR manager can modify and complement this competency dictionary.

The HR manager then entered the job position "Programmer" and the keywords that are shown in Table 1 to identify other appropriate competencies in the datamining module.

Table 1. Keywords used for job position "Programmer"

Keyword
CNC
Java
NET
Postgres

In the next step, the request for searching job vacancies with the title "Programmer" and the inserted keywords is created. The Datamining Module then automatically searched the www.jobs.com work portal and excluded irrelevant job vacancies. The statistics on the number of job vacancies used and excluded are shown in Table 2.

Table 2. The number of used and excluded job vacancies (job vacancies in a non-Czech language or the absence of any of the keywords)

Type	Number
Excluded job vacancies	57
Used job vacancies	63
$\sum$	120

Then, the module suggests competencies that exist in the system database for other job positions and also occur in found job vacancies. Table 3 shows selected proposed competencies.

Table 3. Proposed competencies existing in the database (not selected for the current position) and found in job vacancies

Competency name	Number of occurrences
PostgreSQL	68
Oracle	28
Office	26

In the next step, the module proposes to include the competences found in the job position, which are in the competency dictionary, but are not filled in the system at any job position. Table 4 shows the most numerous competencies.

Table 4. Competencies that are listed in the competency dictionary but are not in the database for any job positions and were found in job positions (10 most numerous)

Competency name	Number of occurrences
java	112
data	36

javascript	31
data (different form)	20
cnc	14
database	11
angular	11
database (different form)	5
ajax	5
android	5

Then, the module shows a list of competencies that are inserted to the current job position in the system but were not found in job vacancies. These competencies are shown in Table 5. The HR manager can decide about removing these competencies from the current job position.

Table 5. A list of competencies that are inserted in the current position but not found in the job vacancies

Competency name
Working with computer
Communicativeness

The, the module shows the statistics of the most common keywords that were found in job vacancies. This keyword list can be beneficial to HR manager because it displays the most used keywords in relevant job vacancies for the current job position.

Table 6. The most used keywords found in job vacancies

Competency name	Number of occurrences
team	40
work	39
holidays	33
development	32
ambience	32
java	24
sql	21
spring	13
documentation	12
hibernate	10

4 Conclusion

The paper proposes a data mining approach for creating a job position in the system for evaluating competencies. The main aim of the data mining approach is to retrieve and process the content of relevant job advertisements on the selected job portal. The output of the approach is a list of relevant words found along with their occurrence in crawled job vacancies that indicate important

IOP Conf. Series: Journal of Physics: Conf. Series **1195** (2019) 012005 doi:10.1088/1742-6596/1195/1/012005

competencies that are required for a given job position. The HR manager then obtains a list of relevant competencies for the selected job position and the selected competencies can add to the profile of job position in the system. Additionally, for the HR manager, the output is also a proposal to remove competencies from the job position profile, because they are not relevant (not found in searched job vacancies). The proposed approach was verified on a specific example and the results were presented.

Acknowledgments
This work was supported within the framework of the IT4Innovations Centre of Excellence project, reg. no. CZ.1.05/1.1.00/02.0070 supported by Operational Programme “Research and Development for Innovations” funded by Structural Funds of the European Union and state budget of the Czech Republic. The work was created during the completion of a Student Grant SGS02/UVAFM/2018 with student participation, supported by the Czech Ministry of Education, Youth and Sports.

References

[1] J. S. Shippmann, et al., "The practice of competency modelling", Personnel psychology. 2000, Volume 53, Issue 3, pp. 703-740.
[2] M. Harzallah, F. Vernadat, "IT-based competency modelling and management: from theory to practice in enterprise engineering and operations", Computers in industry, 2002, Volume 48, Issue 2, pp. 157-179.
[3] M. A. Campion, et al., "Doing competencies well: Best practices in competency modelling" Personnel Psychology, 2011, Volume 64, Issue 1, pp. 225-262.
[4] E. García-Barriocanal, M. A. Sicilia, S. Sánchez-Alonso, "Computing with competencies: Modelling organizational capacities. Expert Systems with Applications", 2012, Volume 39, Issue 16, pp. 12310-12318.
[5] B. Walek, O. Pektor, R. Farana, “Proposal of the Web Application for Selection of Suitable Job Applicants Using Expert System”, Software Engineering Perspectives and Application in Intelligent Systems. Springer, Cham, 2016, pp. 363-373.
[6] O. Pektor, B. Walek, I. Martinik, “Proposal of a system for evaluating competencies”, Proceedings of the 2018 19th International Carpathian Control Conference, Szilvasvarad, 2018, pp. 502 – 507.
[7] K. D. Jones, “Fuzzystring library”, https://github.com/kdjones/fuzzystring

IOP Conf. Series: Journal of Physics: Conf. Series **1195** (2019) 012006 doi:10.1088/1742-6596/1195/1/012006

Defect detection in textile fabrics with snake active contour and support vector machines

Prachya Bumrungkun

Dept. of Electrical Technology, Faculty of Agriculture and Technology Rajamangala University of Technology Isan Surin Campus Surin, Thailand
Email: prachaya.bum1198@gmail.com

Abstract. This paper presents the automatic fabric defect detection in the textile manufacturing industry. The method utilizes edge detection via Snake active contour models to extract the fabric feature. The two-dimensional principal component analysis is employed for data reduction. Then, the outcome of the data reduction passes through the support vector machine as the classifier for classification. The performance of the method is evaluated with the 900 – fabric images for efficiency and effectiveness. The experimental results show that the method is able to detect the defection on the textile with the average accuracy of 98.77 %. The results indicate that the proposed method is outstanding on the defect detection of the textile.

1. Introduction
Nowadays, the textile manufacturing industry employs the machines to handle the job for the large volume and rapidity. Eventually, the defection on the fabric could occur due to the error of the machine. The inspection must be performed instantly to detect the defection to prevent further damage on the production line. The defect detection to spot the defect fabric is necessary on quality control of the process. In the past, the defect detection utilized simple process via visual inspection. The overall accuracy yielded no more than 80% [1] .Thus, many researches and developments have continuously been explored on the new techniques and different methods issue for the high effectiveness and more accurate. Out of these various passes, the principle of image processing techniques [2], [3] give promising result of the effectiveness and accuracy on the detection since the computer, instead of the human being, is employed to make the decision on the defect.

2. Literature review
The fabric defect detection approaches can largely be classified into 4 groups; i.e. statistical, spectral, model based, and structural approaches. Some of the previous works on statistical approach utilized Fisher criterion based denoising auto encoding [4]. The obtained results have high detection accuracy even with insufficient negative samples but time consume. Some utilized Genetic algorithm with homogeneity, contrast, correlation, hue, and saturation to measure the performance [5]. The method needed very large number of high quality images. The Kernal principal component analysis [6] was also employed. The measurements were on the detection success rate, specificity, and sensitivity. The method high true detection rate and low cost for online fabric inspection. Also artificial neural back propagation technique was introduced with 5 features as the feature input [7]. The efficiency of the model was reliable to identify the defect but not appropriate for all types of defects. Another worked with the pulse coupled neural network with mean and standard deviation to measure the performance [8], utilizing embedded system for less computation.

Some on spectral approach utilized 2D Fast Fourier transform with several measurements [9]. The method determined yarn paths even in critical circumstance. The Raden signature and Discrete wavelet transform [10] was introduced with variance, uniformity, and entropy to measure. The method applied to pattern less fabrics and patterned fabrics. Some utilized the discrete cosine transform [11] which would be extended for colour image. On the model based approach, some applied Fuzzy clustering [12] with blind weave detection which need no prior knowledge of the fabric. Also, Neuro Fuzzy classifier was employed with spatial and spectral features [13] but the method needed large data samples to improve the accuracy. Some introduced multi resolution combined statistical and spatial frequency with various measurements [14]. The method yielded good success rate of classification. On the structural approach, some utilized canny edge detection [15] which have fast detection and extraction on the process. The other employed discrete wavelet transform [16]. The method was able to label the boundaries of the defective areas.

This paper interests on structural approach utilizing snake active contour models for defect detection of the textile with the two-dimensional principal component analysis for the data reduction as the tools for feature extraction and the support vector machine as the classifier for classification. The objectives are on analysing the local fabric textile in physical dimension or in general of the textile, to increase the accuracy with efficiency and effectiveness on the defect detection of the textile, to increase the effective productive in the line of product, increase quality of the outputs, reduce time consumption in justification on the outputs and reduce cost of operation via advance technology.

This paper is organized into five sections, including this introduction and literature review sections. Section 3 presents all processed of defect detection method and data. The experimental results are shown in section 4. Finally, the conclusion of this paper are given in section 5.

3. Data and method

3.1. Data

The data in this work consisted of 5 defection groups of the textile materials, each with 50 figures, total of 250 images. Each image has the 256*256 pixels on the gray scale. Each group are as follows: Group A – Gout, Group B – Knot, Group C – Burl, Group D - Warp float, and Group E - Big knot. Figure 1 shows the defection on each group.

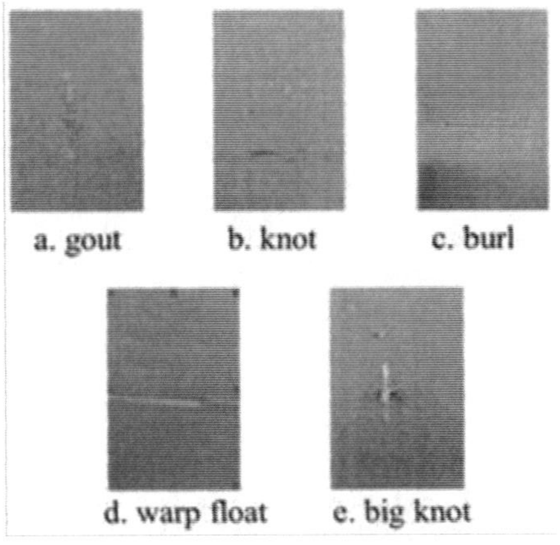

Figure 1. The defection of the fabric a. gout, b. knot, c. burl, d. warp float, and e. big knot.

3.2. Method

The proposed method is shown in figure 2 and the processes are separated into 2 groups; i.e. feature preparation and classification. The first group contained 2 units; i.e. feature extraction and data reduction. First, the feature of the image is extracted via the Snake Active Contour Models. Then, the

data reduction is follow, utilizing 2DPCA to reduce size of the image for training and testing, characteristic of the image still preserved on the process. Finally, the SVM is employed as the classifier for classification.

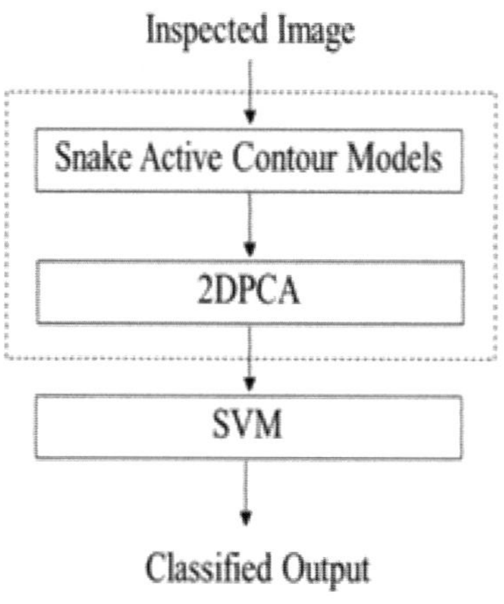

Figure 2. The proposed method.

The proposed method is as follow;

1) Inspected image

The inspected image is the image of 256*256 pixels on gray scale level as the input to the system for feature extraction.

2) Feature Preparation

Feature preparation on the top box in figure 2 consists of 2 units; i.e. feature extraction with Snake active contour model and data reduction with 2DPCA.

a) Snake active contour model [17]: A snake falls into closet local energy minimum. The local minimum of the snake energy comprise the set of alternative solutions. A higher level knowledge is needed to choose the correct one from these solution, High-level reasoning and user interaction. These high-level methods can interact with the contour model by pushing it toward an appropriate local minimum. They rely on other mechanisms to place them near the desired contour. The existence of such an initializer is application dependent. Even in the case of manual initialization, snakes are quite powerful in refining the user's input. Parametric representation: v(s) = (x(s), y(s)). Snake Energy is representation in eq.1

$$E_{snake} = \int E_{\text{int}}(v(s)) + E_{image}(v(s)) + E_{con}(v(s))ds \tag{1}$$

Where Eint = internal energy due to bending; Eimage = image forces pushing the snake toward image features; Econ = external constants are responsible for putting the snake near the desired local minimum. The snake is a controlled continuity spline.

b) 2DPCA is only to eliminate the relevance of the image column and ignore the relevance of image in row [18]. The first one is in horizontal direction and the second is in vertical direction. The process of 2DPCA is described as follow: given image A, the feature matrix B is obtain by the compression of 2D image in horizontal direction in eq.2

$$B = AU \tag{2}$$

Then, transpose B and input BT into 2DPCA, and determine the transform matrix V. Project BT onto V in eq.3.

$$CT=BTV \tag{3}$$

The result Feature matrix is shown in eq.4

$$C=VTB \tag{4}$$

3) Classification

If features are extracted well, then any simple classifier is appropriate. SVM is a supervised machine learning paradigm capable of solving linear and non-linear classification and regression problems. SVM generates input-output mapping functions from a set of labeled training data. The goal of SVM is to produce a model (based on the training data) which classifies the test data. A classification task usually involves separating data into training and testing sets. SVM can classify data separated by non-linear and linear boundaries. SVM maps input vectors to a higher dimensional space where a maximal separating hyperplane is constructed. Two parallel hyperplanes are constructed on each side of the hyperplane that maximizes the distance between the two parallel hyperplanes. An assumption is made that the larger the margin or distance between these parallel hyperplanes the better the generalization error of the classifier [19].

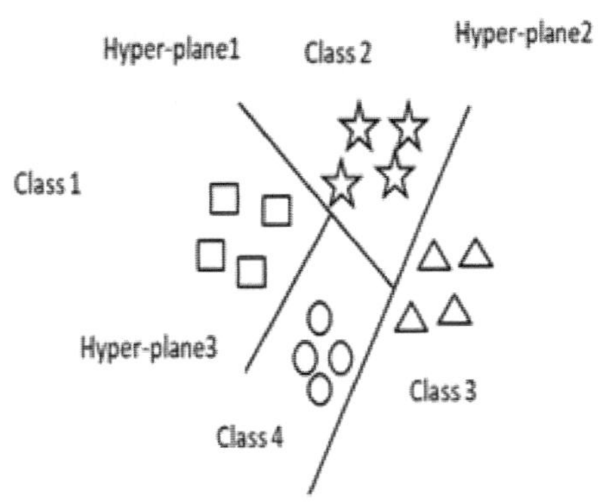

Figure 3. Group of the defect image SVM classification.

4. Experiments and experimental results

We classified images into two categories: defect and non-defect. We used 50 data to train and another 100 data to test each dataset. Additionally, we employed a three-fold cross-validation criterion to train and test. The performance of the classifier was evaluated based on the calculation of the following accuracy.

Table 1. The results of the proposed method.

Image Prototype	Classification detection by SVM		
	Defect	Non- defect	% Accuracy
Non-defect	0	150	100
Gout	148	2	98.67
Knot	147	3	98.00
Burl	148	2	98.67
Warp float	147	3	98.00
Big knot	149	1	99.33
Average			98.77

IOP Conf. Series: Journal of Physics: Conf. Series 1195 (2019) 012006 doi:10.1088/1742-6596/1195/1/012006

Table 1 shows the detect defection efficiency of the textile. Utilizing the combination of structural approach with Snake active contour model for feature extraction and 2DPCA for data reduction as the feature preparation and support vector machine as the classifier for classification, the average accuracy on the classification of 98.77% is achieved. Example of defect and non-defect images shown in figure 4.

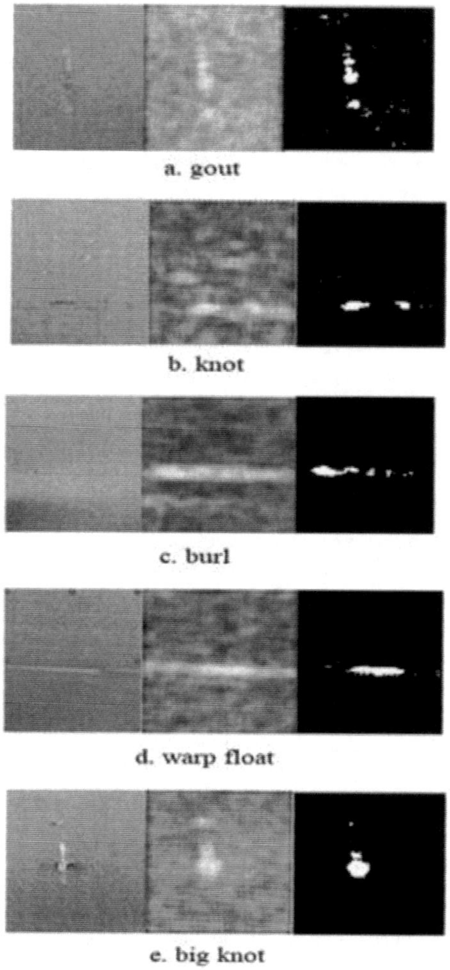

Figure 4. Example of defect and non-defect images.

5. Conclusion

The experimental result indicates that the proposed method, which composed of the combined process of the edge detection with snake active contour model and 2DPCA for feature preparation and support vector machine as the classifier for classification, yields an outstanding average accuracy on justification the defect textile with the appropriate data size on the output from 2DPCA. This process preserved characteristic of the materials and had less time consuming on the training and testing. However, the amount of illumination still create some problem. In this experiment, the limitation are as follows; the appropriate light condition is on day time only and also need appropriate distance on focusing the textile. However, the method is able to apply in other areas but need the amount of data large enough for training and testing on each group.

References

[1] R.Chin, "Automated visual inspection techniques and applications : A bibliography," Pattern Recognition15(4), pp.343-357,1982.
[2] M. Fathu Nisha, P.Vasuki, and S. Mohamed Monsoor Roomi, "Survey on various defect detection and classification methods in fabric images", J.Environ.Nanotechnol., Vol.6 No.2,pp.20-29, 2017.
[3] B. Mingde, S. Zhigang and L. Yesong, "Textural fabric defect detection using adaptive quantized gray-level co-cocarrence matrix and support vector description data," *Information Technology Journal*, Vol.11, No.6, pp.673-685, 2012.
[4] Yundong Li, Weigang Zhao and Jiahao Pan, "Deformable patterned fabric defect detection with fisher criterion based deep learning", *IEEE Trans. on Automation Sci. Engg.*, pp. 01-09, 2016.
[5] P. Dakhole, S. kalode and A.Patel, "Fabricfault detection using image processing matlab," *Int. J. Emerging Trends Engg. Management Res.,* 2(1), 2016.
[6] Junfeng Jing, Xiaoting Fan and Penegfei Li, "Automated fabric defect detection based on multiple gabor filters and KPCA", *Int. J. of Multimedia and Ubiquitous Eng.*, 11(6), pp.93-106,2016.doi:10.14257 /ijmue.2016.11.6.09
[7] Gangandeep Singh, Gurpadam Singh and Mandeep Kaur, Fabric Defect Detection using series of image Processing Algorithm & ANN operation, *Global J., Comp. Tech.*, 4(2), pp. 225-228, 2016.
[8] Tiangpeng Feng, Lian Zou, Jia Yan, Wenxuan Shi, Yifeng Liu, Cien Fan and Dexiang Deng, Real time fabric defect detection using accelerated small scale over completed dictionary of sparse coding, *Int. J. Advanced Robotic Systems.*, 13(1), 01-09(2016). doi: 10.5772/62058
[9] Nasira and Banumathi, P. M., Plain woven fabric defect detection based on image processing and artificial neural networks, *Int. J. Comp. Trends Tech.*, 6(4), 226-229(2013).
[10] Dandan Zhu, Ruru Pan and Wiedong Gao, Yarn dyed fabric defect detection based on autocorrelation function and GLCM, *AUTEX Research J.*, 15(3), 226-232(2015). doi:10.1515/aut-2015-0001
[11] Ali Rebhi, SabeurAbid and FarhatFraeich, Texture defect detection using local homogeneity and discrete cosine transform, World Applied sciences Journal, 31(9), 1677-1683(2014). doi:10.5829/idosi.wasj.2014.31.09.156
[12] Schneider, and Dorit Merhof, D., Blind Weave Detection for Woven fabrics, Springer, Dorian Pattern Anal. Appl., 725-737 (2014).
[13] Eldessouki, M., Hassan, M., Qashqari, K. and Shady, E., Application of principal component analysis to boost the performance of an automated fabric fault detector and classifier, Fibers and Textiles in Eastern Europe, 51-57(2014).
[14] Sabeenian, R. S., Paramasivam, M. E. and Dinesh, P. M., Computer vision based defect detection and identification in handloom silk fabrics, Int. J. of Comp. Appl., 42(17), 41-48(2012).
[15] Halimi Abdellah, Roukhe Ahmed and Ouhmad Slimane, Defect detection and identification in textile fabric by SVM Method, IOSR J. Eng., 4(12), 69-77(2014).
[16] Ibrahim Celik, Cannan Dulger and Mehmet Topalbekiroglu, L., Fabric defect detection using linear filtering and morphological operations, Indian J. Fibre and Textile Research, 39(3), 254-259(2014).
[17] M. Klass, A. Witkin, and D. Terzopoulos, "Snake: Active countour models," Internationanl Journal of Computer Vision, pp. 321-331,1988.
[18] A. Srikaew, K. Attakikitmoncol and W. Kidsang.," Detection of Defect in Textile Fabrics using Optimal Gabor Wavelet Network and Two-Dimonsional PCA," Suranaree Journal of Science and Technology, Vol.18,no.1, pp 436-445, 2011.
[19] Krebel, U., "Pairwise Classification and Support Vector Machines. Advances in Kernel Methods -Support Vector Learning", Cambridge, MA, MIT, pp. 255-268, 1999.

IOP Conf. Series: Journal of Physics: Conf. Series **1195** (2019) 012007 doi:10.1088/1742-6596/1195/1/012007

Classification of pulmonary tuberculosis lesion with convolutional neural networks

T Karnkawinpong and Y Limpiyakorn

Department of Computer Engineering, Chulalongkorn University, 254 Phayathai Road, Pathumwan, Bangkok 10330, Thailand
Yachai.L@chula.ac.th

Abstract. The concept of computer-aided diagnosis (CAD) for chest x-rays (CXR) has been around for the past fifty years. CAD can help in early diagnosis and reduce the deaths caused by late diagnosis and lack of treatment. Applying deep learning techniques for classification of medical images has seen considerable growth in recent years. Convolutional Neural Networks (CNNs) are a class of powerful generative models well known for image classification and segmentation. This paper has studied three deep neural networks: AlexNet, VGG-16 and CapsNet, for classifying tuberculosis in CXR images. The customized models are created using the datasets acquired from National Library of Medicine and private Thai datasets. Data augmentation with shuffle sampling is used to prevent overfitting in the constructed models. The performance of classifiers has been evaluated with the measures: accuracy, sensitivity and specificity. All model accuracy increases with the augmented dataset. The method of affine transformation has also applied to investigate the model accuracy when predicting the test set contains variant instances unseen in the training CXR images.

1. Introduction

Tuberculosis remains the infectious disease that causes deaths every year worldwide. TB is mostly found in the lung region that is termed as Pulmonary Tuberculosis (PTB). PTB commonly manifests upper lung and it is hard to confirm its presence as it appears different pathological patterns depending on various factors. PTB is a curable disease. However, due to the massive workload of physicians, it inevitably results in long and deleterious time-to-treatment periods for patients. A computer-aided triage system would mitigate these issues by providing the initial CXR interpretation to output a classification of normal/ infected per image.

The concept of computer-aided diagnosis for chest x-rays has been around for the past fifty years. Traditionally, CAD systems use machine learning techniques to perform tasks by analysing relationships of existing data. In this study, we have studied deep neural networks, a type of deep learning that employs multiple hidden layers and has been remarkably successful for image classification. The technique is applied for detecting TB lesions on the CXR images. The proposed approach is promising due to its ability to excel with high-dimensional datasets, such as images. Convolutional Neural Networks are a class of powerful generative models with a variety of architectures. They overcome old method performance of image classification without segmentation and manual feature selection. Features are automatically extracted via their own architectures. In literature, researchers have proposed various CNN classifiers such as AlexNet in 2012 [1], VGG in 2014 [2], Inception V3 in 2015 [3], ResNet in 2015 [4], and Xception in 2016 [5]. To overcome the performance of image classification, the latest CNN architecture was proposed by Sabour et al. in

2017, so called capsules or capsules network (CapsNet) [6]. The classifier achieved state of the art performance on MNIST and it is claimed as the best-performing classifier for predicting data with affine transformations that are excluded from training datasets. In this work, the performance of three different deep neural networks were evaluated: AlexNet, VGG-16 and CapsNet for classifying TB on CXR images.

2. Related work

In the world of CXR classification of TB, the first CNN model for TB detection is AlexNet proposed by Hwang et al. in 2016 [7]. In the model, pre-trained was used as the upper layer except in some lower layers where fine tuning was performed. Montgomery and Shenzhen datasets were used as testing data, whereas the model was trained and validated on customized dataset. The AUC performances on test datasets were 0.93 and 0.88 respectively. Cao et al. [8] applied GoogLeNet [9] model for tuberculosis diagnostics on mobile. They used pre-trained model and achieved the accuracy of 89.6%. Hooda et al. [10] used customized CNNs based on LeNet and AlexNet architectures for detection of TB. They used CXR images from Montgomery and Shenzhen datasets. The customized CNNs had 19 layers and consisted of 7 Conv layers, 7 ReLu-layers and 3 Fully-connected layers. Dropout layers were used for preventing the overfitting layer. They compared performance of three different optimizers and found that Adam optimizer overcome the others achieving the accuracy of 94.73% and validation accuracy of 82.09%. Liu et al. [11] proposed TB detection model based on AlexNet and GoogLeNet architectures with different model parameters. The model was trained with a very unbalanced dataset. The shuffle sampling technique was used to augment data resulting in improving the accuracy of AlexNet from 53.02% to 85.68% and improving that of GoogLeNet from 56.11% to 91.72%. Another study by Rajaraman et al. [12] on comparison of deep learning models with CXRs aimed to compare the performance of their customized CNN with five pre-trained CNNs (AlexNet, VGG-16, VGG-19, Xception and ResNet) using Montgomery, Shenzhen, Kenya and India datasets, maintained by National Library of Medicine (NLM) and National Institutes of Health (NIH). The customized CNN model for the binary classifying as normal or TB, consisted of 3 Conv layers, 3 max pooling layers, 1 dropout layer and 2 Fully-connected layers. They used softmax classifier with SGD momentum and L2-regularization optimizer. Each Conv layer was followed by batch normalization [13] and ReLu layer. The performance of customized model achieved accuracy of 0.824 and AUC of 0.900. The comparison results reported that without optimal features, AlexNet achieved the highest accuracy and AUC on all the four datasets. The datasets were randomly split into 80% for training and 20% for testing. However, with optimal features, VGG-16 outperformed the other models.

3. Dataset

Only chest radiographic images of normal and TB lungs were collected from the datasets of National Library of Medicine (NLM) and the Ministry of Public Health, Thailand. All the CXR images contained in four private Thai datasets are deidentified and each of them is confirmed by a radiologist to ensure all images are labelled accurately. The dataset contains totally 1390 images: normal lung 701 images (example as shown in figure 1) and having TB lesion 689 images (example as shown in figure 2). The class label is assigned to value "1" representing TB positive, or "0" denoting TB negative or normal. All images were down-sampled size to 128x128 pixel resolution to suit the input requirements for the customized model.

The method of shuffle sampling is applied for increasing data to prevent overfitting in the constructed models. The size of dataset was increased from 1390 to 1986 images separated into training data of 1588 images (80%), and test data of 398 images (20%). For further investigation of the model performance, another larger dataset was also created using shuffle sampling. The size of dataset was boosted to 3310 images separated into 80% training data (2648 image), and 20% test data (662 images).

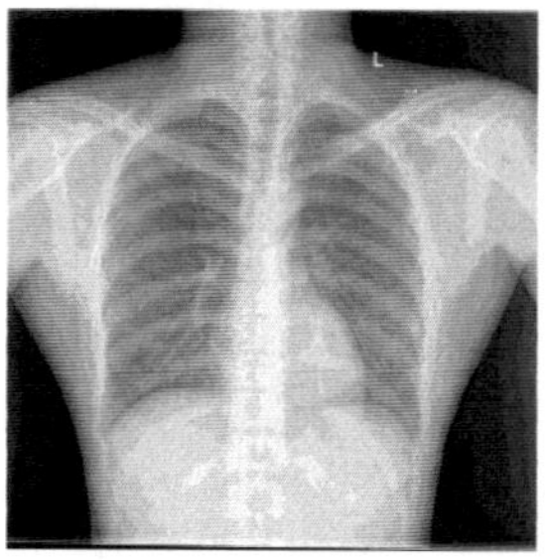

Figure 1. Normal CXR image.

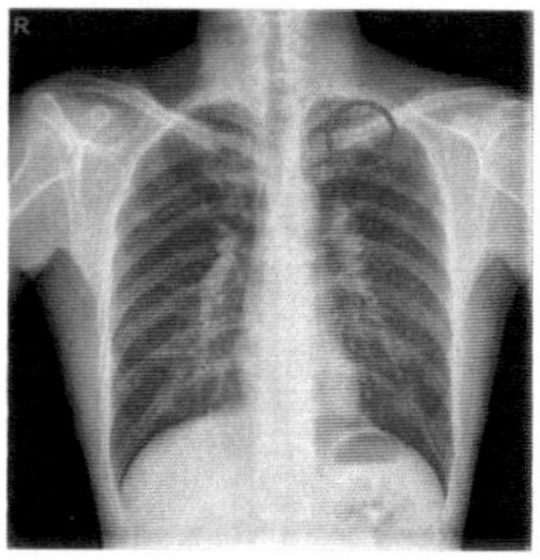

Figure 2. TB CXR image.

4. Architecture

In this work, we use a Windows system with Intel(R) Core (TM) i7-6700k CPU @ 4.00GHz, 8 GB RAM, a Nvidia GTX 960 4 GB graphical processing unit (GPU), Keras with Tensorflow backend, and CUDA 8.0 for GPU acceleration.

We adopted and customized the AlexNet architecture [12] by reducing the number of filters and neurons in fully connected layers for making the structure simpler, that is, it requires less time and fewer number of parameters computed. The customized architecture contains a sequential, 8-layered CNN for classification as TB or normal. It consists of 5 convolution layers (Conv layer), 3 fully-connected layers (FC layer) and 2 dropout layers. The learning rate is 0.001 and Adam [14] is selected as optimizer algorithm.

The customized model of VGG-16 contains a sequential, 19-layered CNN: 16 convolution layer, 3 fully-connected layers and 2 dropout layers. The learning rate is 0.1, the value of learning rate decay is 0.0001, and the optimizer algorithm is SGD.

The CapsNet architecture presented in [6] is applied in this work. The structure is shallow with 2 convolution layers (traditional convolution layer and primary capsules layer) and 1 fully-connected layer (DigitCaps layer). Conv1 is the layer used for extracting local features that are then used as inputs to the primary capsules (PrimaryCaps). The second layer or PrimaryCaps is a convolution capsule layer with 4 channels. Each primary capsule consists of 8 convolution unit (8D capsules) with a 3x3 kernel and a stride of 2.

5. Evaluation

5.1. Evaluation metrics

The performance of the classifiers is evaluated by three measures: 1) accuracy, 2) sensitivity (also known as recall or the true positives rate) and 3) specificity (also known as false positive rate).
Given that TP denotes the number of true positives; FP denotes the number of false positives; TN denotes the number of true negatives; and FN denotes the number of false negatives. Sensitivity, Specificity, and Accuracy are defined as in equation (1), (2), and (3), respectively.

$$Sensitivity = \frac{TP}{TP + FN} \tag{1}$$

$$Specificity = \frac{TN}{FP + TN} \tag{2}$$

$$Accuracy = \frac{TP + TN}{TP + FP + TN + FN} \tag{3}$$

5.2. Comparison of accuracy of three models

For AlexNet, the training images are divided into 128 batches and the model is trained for 200 epochs. We randomly initialized weights in each layer. Biases were also random. After each epoch, weights were updated by Adam optimizer [14]. We performed tests using different learning rates and found the

best performance learning rate is 0.001. AlexNet achieved the highest accuracy of 92.90% on shuffle sampling size of 3310.

Regarding VGG-16, the training images were divided into 128 batches and the model was trained for 200 epochs. The weights in each layer were randomly initialized. Biases were also random. After each epoch, weights were updated by SGD optimizer. The learning rate is 0.1, and learning rate decay is 0.0001. VGG-16 defeated AlexNet and achieved the highest accuracy of 94.56% on shuffle sampling size of 3310.

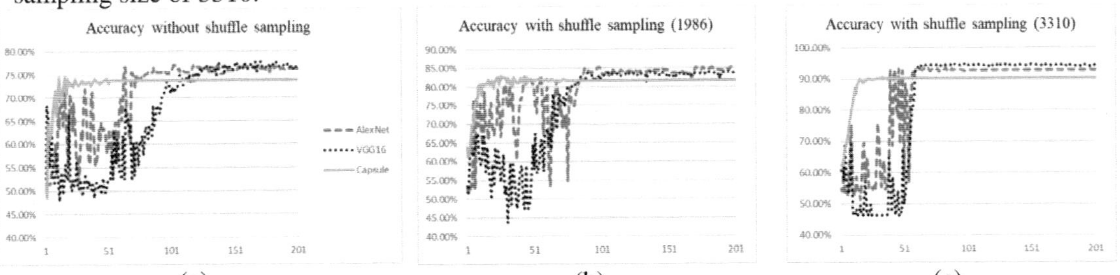

(a) (b) (c)

Figure 3. Comparison of three model accuracy with (a) non-shuffle sampling. (b) shuffle sampling size of 1986. (c) shuffle sampling size of 3310.

For CapsNet, the training images were divided into 64 batches and the model was trained for 200 epochs. The weights in each layer were randomly initialized. Biases were also random. After each epoch, weights were updated by margin loss function [6] and Adam optimizer [14]. The learning rate is 0.001. The best accuracy of CapsNet yielded 90.33% on shuffle sampling size of 3310. Compared to VGG-16 and AlexNet, CapsNet was defeated. The comparisons of the accuracy among three models using non-shuffle sampling, shuffle sampling sizing 1986 and 3310 are visualized in figure 3 (a), (b), (c), respectively.

Summary of all the measures of all classifiers is detailed in table 1, including the values of accuracy, sensitivity and specificity. All classifiers achieve higher accuracy, sensitivity, and specificity on larger datasets. The value of sensitivity is lower than specificity for all models on all datasets. Currently, the highest sensitivity and specificity provided by VGG-16 on shuffle sampling (3310) are 92.83% and 96.06%, respectively.

Table 1. Summary of performance of the three customized models.

	Non-shuffle sampling			Shuffle sampling (1986)			Shuffle sampling (3310)		
Performance	AlexNet	VGG-16	CapsNet	AlexNet	VGG-16	CapsNet	AlexNet	VGG-16	CapsNet
Accuracy	75.18%	**76.26%**	73.74%	**83.92%**	81.91%	82.16%	92.90%	**94.56%**	90.33%
Sensitivity	73.33%	**74.07%**	72.59%	**82.45%**	80.42%	79.37%	92.51%	**92.83%**	86.32%
Specificity	76.92%	**78.32%**	74.83%	**85.40%**	82.30%	84.69%	93.24%	**96.06%**	93.80%

5.3. Affine transformation

Practically, CXR images are not truly vertical, i.e. there is variance in slightly skew and rotation. To investigate the performance of the three classifiers on the dataset containing affine transformations of CXR, the three models were trained using the shuffle sampling dataset of size 3310 without any affine transformation instances. The constructed three models were then tested with the test set containing 300 normal images and 300 affine transformed images, created with slightly random angle from random data in non-shuffle sampling dataset. The result reported that CapsNet outperformed the others with the accuracy of 68.50%, compared to that of 60.33% of AlexNet and that of 58.50% of VGG-16.

6. Conclusion

We evaluate the efficacy of CapsNet for classifying TB in CXR images, compared to AlexNet and VGG-16. The datasets acquired from NLM and the Ministry of Public Health, Thailand were augmented with the shuffle sampling technique. The results show that the accuracy of all the three models is higher compared to that of non-shuffle sampling. Augmenting the dataset could further increase the model accuracy. With shuffle sampling of size 1986, the best-performing classifier is Alexnet, whereas VGG-16 outperforms the other two classifiers on shuffle sampling of size 3310 and non-shuffle sampling datasets. Compared to Alexnet and VGG-16, CapsNet is insignificantly defeated, its training epochs are much less than, though. Moreover, CapsNet achieves the highest accuracy when testing with the dataset that contains variant instances generated by affine transformation. Currently, with shuffle sampling of size 3310, the VGG-16 classifier achieves the highest sensitivity of 92.83%, and the highest specificity of 96.06%. We believe that deep neural networks could accurately classify TB in CXR images. In order to provide valuable information to significantly decrease time-to-diagnosis, further improvement on achieving extremely high sensitivity (confidently classify as having TB lesion) is desired as well as the improvement of the model so that it could accurately predict the variant and unseen images.

References

[1] Krizhevsky A, Sutskever I and Hinton G E 2012 ImageNet classification with deep convolutional neural networks. In: *Proceedings of the 25th International Conference on Neural Information Processing Systems - Volume 1,* (Lake Tahoe, Nevada: Curran Associates Inc.) pp 1097-105

[2] Simonyan K and Zisserman A 2014 *Very Deep Convolutional Networks for Large-Scale Image Recognition*

[3] Szegedy C, Vanhoucke V, Ioffe S, Shlens J and Wojna Z B 2016 *Rethinking the Inception Architecture for Computer Vision*

[4] He K, Zhang X, Ren S and Sun J 2016 Deep Residual Learning for Image Recognition. In: *2016 IEEE Conference on Computer Vision and Pattern Recognition (CVPR),* pp 770-8

[5] Chollet F 2017 *Xception: Deep Learning with Depthwise Separable Convolutions*

[6] Sabour S, Frosst N and E Hinton G 2017 *Dynamic Routing Between Capsules*

[7] Hwang S, Kim H-E, Jeong J and Kim H-J 2016 A novel approach for tuberculosis screening based on deep convolutional neural networks. In: *SPIE Medical Imaging*: SPIE) p 8

[8] Cao Y, Liu C, Liu B, Brunette M J, Zhang N, Sun T, Zhang P, Peinado J, Garavito E S, Garcia L L and Curioso W H 2016 Improving Tuberculosis Diagnostics Using Deep Learning and Mobile Health Technologies among Resource-Poor and Marginalized Communities. In: *2016 IEEE First International Conference on Connected Health: Applications, Systems and Engineering Technologies (CHASE),* pp 274-81

[9] Szegedy C, Wei L, Yangqing J, Sermanet P, Reed S, Anguelov D, Erhan D, Vanhoucke V and Rabinovich A 2015 Going deeper with convolutions. In: *2015 IEEE Conference on Computer Vision and Pattern Recognition (CVPR),* pp 1-9

[10] Hooda R, Sofat S, Kaur S, Mittal A and Meriaudeau F 2017 Deep-learning: A potential method for tuberculosis detection using chest radiography. In: *2017 IEEE International Conference on Signal and Image Processing Applications (ICSIPA),* pp 497-502

[11] Liu C, Cao Y, Alcantara M, Liu B, Brunette M, Peinado J and Curioso W 2017 TX-CNN: Detecting tuberculosis in chest X-ray images using convolutional neural network. In: *2017 IEEE International Conference on Image Processing (ICIP),* pp 2314-8

[12] Rajaraman S, Antani S, Candemir S, Xue Z, Kohli M and Thoma G 2018 *Comparing deep learning models for population screening using chest radiography*

[13] Ioffe S and Szegedy C 2015 *Batch Normalization: Accelerating Deep Network Training by Reducing Internal Covariate Shift*

[14] Kingma D and Ba J 2014 *Adam: A Method for Stochastic Optimization*

IOP Conf. Series: Journal of Physics: Conf. Series **1195** (2019) 012008 doi:10.1088/1742-6596/1195/1/012008

Generating Trading Strategies Based on Candlestick Chart Pattern Characteristics

Siriporn Thammakesorn and Ohm Sornil

School of Applied Statistics, National Institute of Development Administration, Bangkok, Thailand

E-mail: siriporn.tha@stu.nida.ac.th

Abstract. Candlestick chart patterns are widely used in stock trading decisions. Patterns of candlestick series are found to provide hints for the price of the next one. This research proposes a technique to generate stock trading strategies employing features which are shown to effectively recognize patterns in candlestick charts. The features are combined into a tree-like trading strategy using the Chi-square Automatic Interaction Detector algorithm. The technique is evaluated using actual stocks from Stock Exchange of Thailand. The results show that the generated strategies are more profitable than other popular trading techniques, such as moving average convergence divergence, exponential moving average, relative strength index, stochastic oscillator and average directional index.

1. Introduction

An investor expects a high return from investing in stocks. Stock prices are affected by many factors, such as economic conditions, political policies, natural disasters, etc. Fundamental analysis and technical analysis are two main approaches which form the basis of most traders' decisions. Fundamental analysis considers economic, political, and detailed studies of companies' financial positions while technical analysis employs indicators based on stock prices and volumes. Candlestick chart [1] is a popular method to display stock price information which consists of open price, high price, low price and close price. Patterns in candlestick charts are found to provide hints for future prices. They were invented in Japan with a long history for trading grains. Nowadays, they are applied to trading other instruments: stocks, futures and currencies. Lu and Shiu [2] study the profitability of candlestick patterns on Taiwan 50 Index component stocks from 2002 to 2009. They find that certain bullish candlestick patterns consistently outperform others. Moreover, they notice that buying signals are generally more effective than selling signals. Goo et al. [3] analyze daily data of 25 component stocks in the Taiwan Top 50 Tracker Fund and Taiwan Mid-Cap 100 Tracker Fund from 1997 to 2006. They find a strong evidence that certain candlestick trading strategies are profitable. In addition, they notice that different candlestick patterns require different holding periods to be profitable. There are more than 60 types of candlestick patterns which make it difficult for human to apply them effectively in investments. Some researchers study the use of candlestick charts and artificial intelligence in stock trading. Izumi, Yamaguchi, Mabu, Hirasawa and Hu [4] create trading rules by combining genetic network programming (GNP) with candlestick charts. It is found that the average profit margin derived from GNP is higher than others. It can be seen that adopting a candlestick chart to create a trading rule can help increase return on investment. Chalothon Chootong [5] create trading signals by combining a neural network (NN) with indicators, price chart patterns, and candlestick chart patterns. It can be seen

IOP Conf. Series: Journal of Physics: Conf. Series **1195** (2019) 012008 doi:10.1088/1742-6596/1195/1/012008

that candlestick chart patterns can help investors make better decisions. Leslie C.O. Tiong, David C.L. Ngo, and Yunli Lee [6] used candlestick patterns to study two AI techniques: artificial neural network (ANN) and support vector machine (SVM) in learning patterns and behaviours of stock prices for future price prediction. It was found that the best technique can predict the stock price correctly with 60 percent accuracy. Our previous research [7] discovers a set of features that can recognize candlestick chart patterns effectively. In this research, a trading strategy is generated from those features using the Chi-square Automatic Interaction Detector (CHAID) algorithm. The goal is to create trading strategies that provide high returns on investment. Experiments are conducted on 10 stocks from Stock Exchange of Thailand over ten years period, and the results are compared with popular trading strategies in details.

In the rest of the paper, the proposed technique is presented including candlestick chart pattern characteristics; results of the experimental evaluations are shown and discussed; and finally conclusions are drawn.

2. Proposed Technique

Our study is shown in Figure 1. Training data is used by the Chi-square Automatic Interaction Detector (CHAID) algorithm to build a trading strategy in form of a classification tree using features which can effectively recognize candlestick chart patterns. The model is then tested with actual price data of unseen periods. This section first introduces candlestick charts, the features used in the proposed method and generation of a trading strategy.

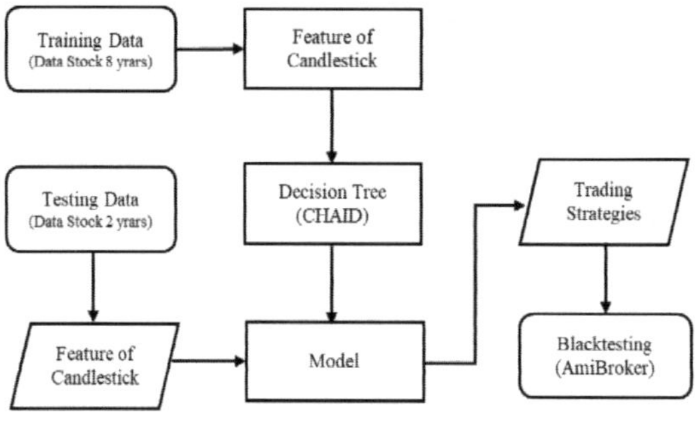

Figure 1. Framework of the research

2.1. Candlestick Chart Characteristics

Candlestick charts [8, 9] originated from Japan. It has a history of more than 200 years for analyzing rice and grain prices. After several decades, the chart is adopted into trading securities and becomes a very popular technical analysis technique. A candlestick (see Figure 2.) shows relationships between open price, high price, low price and close price.

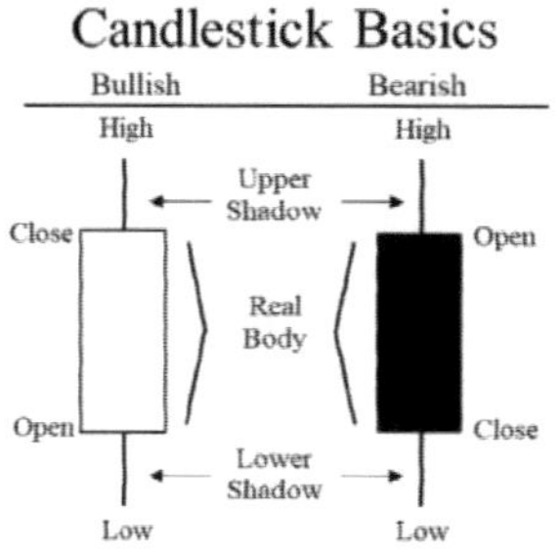

Figure 2. A candlestick

Patterns of candlesticks are found to provide hint on future prices. However, there are more than 60 patterns making it difficult for an investor to use them effectively. In our previous work [8], a set of 26 features for automatic recognition of candlestick chart patterns is proposed and evaluated. The features

are based on colours, sizes and relative positions of candlesticks. We can then make use of those features to construct a trading strategy for predicting future prices. The features consist of:

- Prior Trend
- Colour of the 1st candlestick
- Colour of the 2nd candlestick
- Colour of the 3rd candlestick
- Size of the 1st candlestick normalized by the size of the largest candlestick
- Size of the 2nd candlestick normalized by the size of the largest candlestick
- Size of the 3rd candlestick normalized by the size of the largest candlestick
- The ratio between the sizes of the 2nd and 1st candlesticks
- The ratio between the sizes of the 3rd and 1st candlesticks
- The ratio between the sizes of the 3rd and 2nd candlesticks
- Gap between the 1st and 2nd candlesticks
- Gap between the 1st and 3rd candlesticks
- Gap between the 2nd and 3rd candlesticks
- Real body gap between the 1st and 2nd candlesticks
- Real body gap between the 1st and 3rd candlesticks
- Real body gap between the 2nd and 3rd candlesticks
- Slope between the low price between 1st and 2nd candlesticks
- Slope between the high price between 1st and 2nd candlesticks
- Slope between the low price between 1st and 3rd candlesticks
- Slope between the high price between 1st and 3rd candlesticks
- The real body size of the 1st candlestick normalized by the largest real body size
- The real body size of the 2nd candlestick normalized by the largest real body size
- The real body size of the 3rd candlestick normalized by the largest real body size
- The real body size of the 1st candlestick normalized by its total size
- The real body size of the 2nd candlestick normalized by its total size
- The real body size of the 3rd candlestick normalized by its total size

2.2. Generating Trading Strategy

A trading strategy is a classification tree with 3 output values: Buy, Sell or Hold. The strategy is created by the Chi-square Automatic Interaction Detector (CHAID) algorithm [10]. CHAID uses Pearson's Chi-square test of independence, which tests for an association between two categorical variables, to determine the p-value which is used to select the variables to split a node. The statistics for the chi-Square test of independence is

$$\chi^2 = \sum_{i,j} \frac{(O_{ij} - E_{ij})^2}{E_{ij}} \tag{1}$$

where i is 1, 2, ..., r and j is 1, 2, ..., c , O_{ij} is the observed (actual) frequency value, and E_{ij} is the expected frequency value. CHAID consists of three steps, i.e., merging, splitting and stopping as follow:

- Merging
  This step will calculate a significant test of each categorical independent variable towards the dependent variables and merge most similar categories of independent variables by selecting the smallest significant or the biggest p-value of a category.
- Splitting
  Splitting step separates a node by considering the biggest significant or the smallest p-value of independent variables obtained from the merging process.
- Stopping

IOP Conf. Series: Journal of Physics: Conf. Series **1195** (2019) 012008 doi:10.1088/1742-6596/1195/1/012008

The stopping step checks for terminating the creation and separation of nodes, with the following requirements:

a) If the depth of the tree reaches the maximum depth limit, the data classification process stops.
b) If the size of a node is less than the minimum node size, then the node will stop splitting.

3. Experimental Evaluations

To evaluate the proposed method, data of 10 stocks from Stock Exchange of Thailand (SET) during 2008 to 2017 are used in the experiments. The data are divided into training data (from 2008 to 2015) and testing data (from 2016 to 2017). Amibroker software [11] is used for backtesting and measuring effectiveness of each strategy. Buying and selling orders will be issued at the opening price on the next day after the day yielding the signal. The proposed method is compared with 8 popular trading strategies which are:

- Buy and Hold
- EMA: If EMA (5) crossovers EMA (20) then buy. If EMA (20) crossovers EMA (5) then sell.
- MACD (1): If Fast (12) crossovers Slow (26) then buy. If Slow (26) crossovers Fast (12) then sell.
- MACD (2): If MACD (12,26) crossovers Signal (9) then buy. If Signal (9) crossovers MACD (12,26) then sell.
- RSI: If RSI (14) is less than 30 then buy. If RSI (14) is greater than 70 then sell.
- STO: If Fast (14) crossovers Slow (3) then buy. If Slow (3) crossovers Fast (14) then sell.
- STO: If Slow (14, 3) is greater than 20 then buy. If Slow (14, 3) is less than 80 then sell.
- ADX: If +DI crossovers -DI then buy. If -DI crossovers -DI then sell.

Table 1. Results of Backtesing

Stock	Trading Method	Net Profit (Percent)	Trading Number	Stock	Trading Method	Net Profit (Percent)	Trading Number
ADVANC	Propose Method	**36.04%**	90	**AMARIN**	Propose Method	**8.92%**	69
	Buy and Hold	33.57%	1		Buy and Hold	-16.36%	1
	EMA	20.68%	12		EMA	-14.15%	21
	MACD(1)	-4.88%	7		MACD(1)	-17.69%	8
	MACD(2)	28.14%	15		MACD(2)	-21.21%	22
	RSI	25.78%	3		RSI	-8.38%	2
	STO(1)	32.10%	56		STO(1)	10.96%	56
	STO(2)	15.66%	9		STO(2)	-16.53%	3
	ADX	18.76%	12		ADX	-4.73%	16
AOT	Propose Method	**36.32%**	98	**BBL**	Propose Method	**34.67%**	1
	Buy and Hold	98.25%	1		Buy and Hold	34.67%	1
	EMA	57.73%	11		EMA	14.96%	17
	MACD(1)	57.02%	5		MACD(1)	25.57%	6
	MACD(2)	30.59%	19		MACD(2)	-6.54%	23
	RSI	14.29%	1		RSI	7.19%	1
	STO(1)	21.18%	63		STO(1)	6.78%	58
	STO(2)	8.35%	6		STO(2)	4.49%	7
	ADX	53.48%	19		ADX	4.63%	26
BKI	Propose Method	**8.40%**	80	**CPF**	Propose Method	**5.09%**	111
	Buy and Hold	0.85%	1		Buy and Hold	36.36%	1
	EMA	-4.73%	22		EMA	-0.44%	15
	MACD(1)	-0.77%	11		MACD(1)	11.33%	8
	MACD(2)	-6.22%	23		MACD(2)	-1.41%	19
	RSI	2.59%	2		RSI	-4.50%	1
	STO(1)	-8.34%	67		STO(1)	0.72%	59
	STO(2)	-4.18%	5		STO(2)	-7.29%	5

	ADX	-3.26%	19		ADX	-1.54%	22
NSI	Propose Method	**11.54%**	79	**PTT**	Propose Method	**85.11%**	118
	Buy and Hold	-11.88%	1		Buy and Hold	84.87%	1
	EMA	-18.22%	18		EMA	53.08%	13
	MACD(1)	-12.88%	8		MACD(1)	45.59%	7
	MACD(2)	-31.53%	21		MACD(2)	27.97%	20
	RSI	-5.60%	2		RSI	22.47%	1
	STO(1)	-12.32%	59		STO(1)	26.22%	57
	STO(2)	-2.92%	7		STO(2)	16.41%	7
	ADX	-19.81%	25		ADX	65.19%	21
SCC	Propose Method	**7.56%**	1	**TDEX**	Propose Method	**37.88%**	3
	Buy and Hold	7.56%	1		Buy and Hold	37.88%	1
	EMA	-4.29%	15		EMA	17.16%	16
	MACD(1)	-2.36%	9		MACD(1)	22.45%	7
	MACD(2)	-9.24%	24		MACD(2)	28.29%	20
	RSI	6.26%	2		RSI	16.71%	2
	STO(1)	2.09%	65		STO(1)	30.69%	59
	STO(2)	22.56%	7		STO(2)	6.75%	6
	ADX	-1.01%	23		ADX	21.56%	23

The results of testing with data of 10 stocks for 2 most recent years (2016 to 2017) are shown in Table 1. Net Profit is the net profit achieved after completing all trades in the period, and Max System Drawdown is the largest peak to valley percentage decline experienced while trading in the period. We can see that strategies generated from the proposed technique yield the highest returns or are ranked among the top strategies for stocks that are in an upward trend. Also the generated strategies suffer the lowest losses or are among the top strategies for stocks that are in a downward trend. For some stocks, there may be a very few strategies that yield higher returns than our strategies; but there is no common strategy that generally outperforms our technique. Overall, strategies generated from the proposed technique outperform 8 other comparative techniques which mean that our technique is applicable to trading actual stocks.

4. Conclusion

Candlestick chart is a popular method to display stock price information. Patterns of candlestick series are found to provide hints for the price of the next one. There are so many types of candlestick chart patterns which are difficult to be used effectively. In our work, twenty six features for recognizing candlestick chart patterns are identified. This paper presents a technique to generate effective trading strategies based on those features. A trading strategy is a classification tree generated by the Chi-square Automatic Interaction Detector (CHAID) algorithm. The technique is compared with eight popularly used strategies on 10 stocks from Stock Exchange of Thailand. The results show that the proposed technique generates trading strategies that outperform other strategies, and the proposed technique is applicable to real-world security trading.

5. References

[1] Nison S 2001 *Japanese Candlestick Charting Techniques (2nd ed.)* New Jersey (Pearson Education US).

[2] Lu T-H and Shiu Y-M 2012 Tests for two-day candlestick patterns in the emerging equity market of Taiwan *Emerging Markets Finance & Trade* **48** 41-57.

[3] Goo Y-J, Chen D-H and Chang Y-W 2007 The application of Japanese candlestick trading strategies in Taiwan *Investment Management and Financial Innovations* **4(4)** 49-79.

[4] Izumi Y, Yamaguchi T, Mabu S, Hirasawa K and Hu J 2006 Trading Rules on The Stock Markets Using Genetic Network Programming with Candlestick Chart *2006 IEEE Congress on Evolutionary Computer* **43** 317-322.

[5] Chootong C and Sornil O 2012 Trading Signal Generation Using A Combination of Chart

Patterns and Indicators *International Journal of Computer Science Issues* **9** 202-209.

[6] Tiong L C O, Ngo D C L and Lee Y 2013 Stock Price Prediction Model using Candlestick Pattern Feature *International Journal Of Interactive Digital Media* **1(3)** 58-64.

[7] Dhammathanapatchara T and Sornil O 2015 Automatic Candlestick pattern recognition. *International Journal of Applied Engineering Research* **10** 1833-1836.

[8] Bigalow S W 2011 *Profitable Candlestick Trading: Pinpointing Market Opportunities to Maximize Profits (2nd ed.)* New Jersey (John Wiley & Sons, Inc.).

[9] Corbitt W A 2012 *Candlestick Charting Demystified* New York (McGraw Hill).

[10] Kass G V 1980 An Exploratory Technique for Investigating Large Quantities of Categorical Data *Applied Statistics* **29** 119-127.

[11] Bandy H B 2012 *Introduction to AmiBroker (2nd ed.)* South Dakota (Blue Owl Press, Inc.).

IOP Conf. Series: Journal of Physics: Conf. Series **1195** (2019) 012009 doi:10.1088/1742-6596/1195/1/012009

Thai music emotion recognition based on western music

S Sangnark, M Lertwatechakul and C Benjangkaprasert

Faculty of Engineering, King Mongkut's Institute of Technology Ladkrabang, Bangkok, THAILAND, 10520

kbchawal@kmitl.ac.th

Abstract. Music emotion recognition is the music emotion detected from people's annotations. In this paper, the Thai music was the evaluated set of a system based on western music training settings. By using valence-arousal values, multiple linear regression, k-nearest neighbours to represent the emotional annotations from the music. We used valence and energy(arousal) from Spotify API to the investigated emotion of Thai music. As a result, the Thai music emotion according to the western music criteria could be understood. The highest f-measure of Thai music from multiple linear regression All feature was 41% and the f-measure of western music from multiple linear regression without tempo feature was 51 %, which are very different because All feature in western music is low efficiency than other models.

1. Introduction

Music, one of the arts, expresses the people's mind and emotion. Many people choose the music style that matches their mood just by following "Playlist of emotions" on music streaming platforms. Therefore, many researchers in the scientific field are interested in the music emotion recognition (MER) [1-11] help to understand the emotion of songs for facilitates many listeners. The knowledge of signal processing, machine learning, psychology and musicology must be incorporated. In signal processing, the method described in [1] adopts 5 audio features such as pitch, tempo, tonality, dynamic and timbre to analyse the kinds of emotion. The work [2] presented the emotional analysis by using RMS energy, Mel-frequency cepstral coefficients, zero-crossing rate, fundamental frequency and voice probability, and EEG signal also. In machine learning, there are many classifiers and regression used in categorizing the emotions from music such as support the vector machine [3-4], random forest [2] and regression model [5-6]. There is use knowledge of musicology for the described characteristic of signal and psychology for described emotions by Valence-Arousal model. However, there has no best model according to the individual musical perception. It means that each people perceive different emotion by listening to the same song because the musical emotion depends on many factors. In general, many peoples can understand the feeling of non-native language music because the music interpretation includes many factors. For example, the western music which based on the same genres turns out to be more popular than music from the other regions. The western music dataset was the most used in analysis works [1-6] while then a non-western one was also used in some researches such as the emotional analysis of Sinhala songs was presented in [7] and the popular North Indian songs analysis was presented in [8], but a few studied on Thai songs. Therefore, we selected Thai songs for this work because the Thai language is native language and Thai song is most popular in the region of research. However, many researchers used the same regional songs for development and evaluate the model, but a few studies on different regional songs in development and evaluate the model. The aim of this work is to created western music model and choose best model from western music to predict emotion of Thai songs. We used Spotify API for check emotion of Thai songs. The results show the

IOP Conf. Series: Journal of Physics: Conf. Series **1195** (2019) 012009 doi:10.1088/1742-6596/1195/1/012009

emotion of Thai songs from western music model and best model efficiency of western music and Thai songs have different or same. The paper is organized as follows. The methodologies of music emotion recognition are described in section 2. Experimental results are discussed in section 3 and concluding in section 4.

2. Methodologies

2.1. Dataset

Western music dataset that we selected is MediaEval2013 [9]. It includes 1000 English MP3 songs from freemusicarchive.org. In these files, there were some partial redundancies. After removing redundancies, the remaining dataset contains 744 songs, which are separated into 619 songs for development and 125 songs for evaluation and can be described its emotion on the valence-arousal model. Thai music dataset we selected is my collection. It includes 125 Thai popular songs.

2.2. Valence-Arousal

Valence-arousal [10-11] is a 2D emotion plane. The horizontal axis is valence, which describes positivity and negativity, and the vertical axis is arousal, which describes the excitement and calmness. The value of VA is between -1 and 1. In this work, we only paid attention to 4 different emotion classes; happy, excited, sad and peaceful. Figure 1 shows the emotion on valence-arousal, (a) western music and (b) Thai music is most happy and excited, so the calculated average tempo of Thai music from data processing that almost very high was 126.731 bpm (beats per minute).

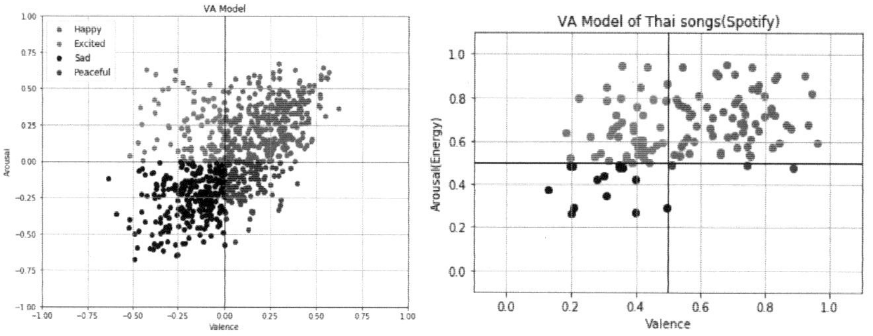

(a) (b)

Figure 1. Emotions on valence-arousal: (a) Western music (b) Thai music.

2.3. Spotify API

We selected valence value from Spotify [12] and used energy value to represent arousal value. The value of valence and energy from Spotify are between 0 and 1, which are different from MediaEval2013. We cannot normalize data because the emotion of songs will be changed, so we plotted the valence and the energy from Spotify to VA model for obtained annotations emotion classes; happy (64 songs), excited (43 songs), sad (15 songs) and peaceful (3 songs) in Figure 1 (b).

2.4. Data processing and feature extraction

Data processing is the important step because the accurate results obtained from quality data. There are many toolboxes that can be used for audio signal analysis. In this work, we relied on the method described in Librosa [13] for extracting features on Python software. These features include pitch, tempo, dynamic and tonality, as presented in table 1.

Table 1. Musical features.

Features	No.	Methods	Description
Pitch	1	Discrete Fourier transforms	The fundamental frequency of the signal is ears perceive or lowest pitch notes.
Dynamic	1	RMS energy	Root-mean-square (RMS) is the computing power of the signal based on the amplitude of the peak value.
Tempo	1	Estimate tempo	Tempo is the pulse at different times of music, which highly effecting on emotion. We extracted the estimated beat of the signal in beat per minutes.
Tonality	12	Chromagram (mean)	Tonality represents the frequency of the musical scale. There are 12 chromagrams related to 12 pitch classes. We assume that these features influence of emotion, so we used 12 features of chromagram included mean chromagram.

2.5. Regression, classification and performance measurement

We used multiple linear regression (MLR), which is a dependent variable correlated with independent variables, to predict the VA values of music and divided into valence and arousal regressions. K-nearest neighbours for classifications which the K values between 1-30 for find out the suitable model of K-NN to predict emotion annotations. In part of performance measurement, we calculated model accuracy by equation (1). In case of the results are not satisfactory and the correct data are not class distribution, it means that the model accuracy is not the best way to reliability. Therefore, we use precision, recall and f-measure by equations (2), (3) and (4), respectively.

$$accuracy = \frac{TP+TN}{TP+TN+FP+FN} \quad (1), \qquad recall = \frac{TP}{TP+FN} \quad (2)$$

$$precision = \frac{TP}{TP+FP} \quad (3), \qquad F-measure = 2\times\left(\frac{precision\times recall}{precision+recall}\right) \quad (4)$$

where TP = True positive, TN = True negative, FP = False positive, FN = False Negative

3. Experimental results

After re-sampling for all music files to 22.5 kHz and mono channels, we selected 15 from 4 music features and divided into valence and arousal regressions. The highest coefficient values of both parts were a dynamic feature. It indicated that the dynamic feature causes most effects on emotion. As well as tempo, which assumed to causes high effects on emotion, resulted in the low coefficient value because the tempo from calculation was not distributed. Table 2 shows the statistically determined R^2 coefficients of valence and arousal values. These were 11.7 % and 52.6 %, respectively. According to the complexity of human emotion, predicting the emotion was more difficult than physical, it was not abnormal in the psychological field for the low R^2 values, especially in terms of emotions. MSE of valence-arousal values were 0.06 and 0.04, respectively. The good result was the minimum value.

Table 2. R-square, Mean square error and P-value.

			P-value		
	R^2	MSE	Pitch	Dynamic	Tempo
Valence	11.7%	0.06	0.437	0.00	0.427
Arousal	52.6%	0.04	0.599	0.00	0.016

IOP Conf. Series: Journal of Physics: Conf. Series **1195** (2019) 012009 doi:10.1088/1742-6596/1195/1/012009

The P-value, especially the pitch in valence model, shown in table 2 is very high. Therefore, the pitch and tempo were reduction feature for investigating the best model accuracy and be used in all case for K-nearest neighbours. The VA values plotted were obtained from evaluating the model on the emotion plane to represented emotion annotations. The highest model accuracy of emotion annotations from the model without tempo features by MLR was 56%, and the highest K-NN model accuracy of emotion annotations from the model without pitch and tempo at K = 14 and K = 15 was 55.2%, as shown in Figure 2.

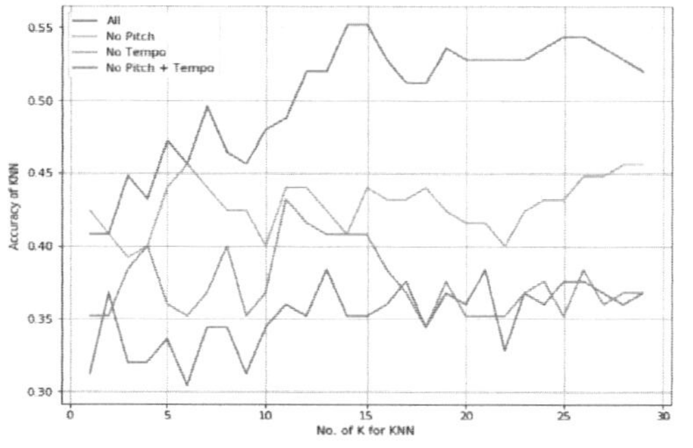

Figure 2. The accuracy of K-nearest neighbours.

Results from MLR and K-NN were satisfied with using to predict emotion. Therefore, we selected all models from MLR and 2 models from K-NN, which are the highest accuracy, to predict the emotion of Thai songs. But all models were selected is not class distribution even if accuracy more than 50%, so we calculated average f-measure, precision and recall to investigating efficient models, as shown in table 3. The highest f-measure from MLR without a pitch and tempo features was 51%, and others f-measure were 50% even if accuracy less than K-NN models at K = 14 and 15. The accuracy and the recall from entire features were more than 50% but the excitement is not accurate, as shown in table 3. The All model from MLR and K-NN was not the best model to predict emotion especially K-NN with All was average lowest accuracy.

Table 3. Accuracy, Precision, Recall, F-measure of MediaEval2013.

	Accuracy calculate	Accuracy	Precision	Recall	F-measure
All[a]	(41+0+19+8)[c] / 125	54.4%	48%	54%	50%
No Pitch[a]	(41+0+20+8)[c] / 125	55.2%	48%	55%	50%
No Tempo[a]	(41+0+21+8)[c] / 125	56.0%	49%	56%	**51%**
No Pitch + Tempo[a]	(40+0+20+8)[c] / 125	54.4%	47%	54%	50%
K = 14[b]	(46+0+18+8)[c] / 125	55.2%	57%	55%	49%
K = 15[b]	(46+0+18+8)[c] / 125	55.2%	57%	55%	49%

[a] Multiple linear regression, [b] K-NN without pitch and tempo, [c] Happy, Excited, Sad, Peaceful.

To predict the emotion of Thai music, 125 popular songs of Thai music collection were selected and then added to the experiment. The property of these songs as same as western music. Then, there were by MLR all model and K-NN no pitch and tempo at K=14 and 15, respectively. The highest f-measure of Thai songs with All model was 41%, as shown in table 4, even though this model in western music was the lowest efficiency. The average of f-measure with entire models less than 50%. The 69% precision of All model was interested because it was only one correct model to predict excited emotion. The sadness from the entire model was not accurate.

Table 4. Accuracy, Precision, Recall, F-measure of Thai music.

	Accuracy calculate	Accuracy	Precision	Recall	F-measure
All[a]	(64+1+2+0)[c] / 125	53.6%	**69%**	54%	**41%**
No Pitch[a]	(64+0+2+0)[c] / 125	52.8%	33%	53%	39%
No Tempo[a]	(63+0+2+0)[c] / 125	52.0%	33%	52%	39%
No Pitch + Tempo[a]	(63+0+3+0)[c] / 125	52.8%	35%	53%	40%
K = 14[b]	(62+0+5+0)[c] / 125	53.6%	34%	54%	40%
K = 15[b]	(62+0+5+0)[c] / 125	53.6%	34%	54%	40%

[a] Multiple linear regression, [b] K-NN without pitch and tempo, [c] Happy, Excited, Sad, and Peaceful.

4. Conclusion

The R^2 values of valence-arousal from All model by MLR in Western music were 11.7% and 52.6%, respectively. To improve the efficiency of the model by P-value. The dynamic was a feature that has the most affects to the emotion. The emotion annotations from VA values without tempo by MLR showed the best accuracy for 56%, and 55.2% by K-NN without pitch and tempo at K = 14 and 15. The highest f-measure from MLR without tempo was 51% and no model suit for used to predict the excited emotion. The All model was low accuracy. We selected 6 models from western music to predict emotion of Thai songs. Even if All model was low accuracy to predict emotion in western music, the results showed that it was highest f-measures 41% and 69% of precision. The results showed that Thai music may be different characteristic from western music, and models from western music have not accurate with excited emotion but most of Thai songs was. Therefore, we cannot use the same model to predict emotions. If someone wants to predict the emotion of regional songs a model for the songs of a specific region should be created.

References

[1] Han K Moo, Zin T and Tun H 2016 *Int. Journal of Scientific & Technology Research vol 5 6* pp 53-56
[2] Zhang F, Meng H and Li M 2016 *Int. Conf. on Natural Computation, Fuzzy Systems and Knowledge Discovery* (Changsha: China) pp 1728-33
[3] Panda R, and Paiva R P 2012 *Proc. of Int. Conf. on Digital Audio* (York: UK)
[4] Lin C, Liu M, Hsiung W and Chang J 2016 *Proc. of Int. Conf. on Machine Learning and Cybernetics* (Jeju: South Korea) pp 375-79
[5] Bai J, Peng J, Shi J and *et al* 2016 *Proc. of Int. Conf. on Cognitive Informatics & Cognitive Computing* (Palo Alto, CA: USA) pp 42-49
[6] Yang Y H, Lin Y C, Su Y F, Chen H H and *et al* 2008 *IEEE Trans. Audio, Speech & Language Processing* vol 16 no 2 pp 448-57
[7] Viraj M G and Jayaratne K L 2016 *Proc. of Int. Conf. on Information and Automation for Sustainability* (Galle: Sri Lanka) pp 1-6
[8] Desmukh S H and Bhirud S G 2014 *Int. Journal of Computer Applications* vol 91 no 4 pp 1-5
[9] Soleymani M, Caro M N, Schimidt E M, Sha C Y and Yang Y H 2013 *Proc. of ACM Int. Workshop Crowdsourcing for Multimedia* (New York: USA) pp 1-6
[10] Russell J A 1980 *J. of Personality and Social Psychology* 39 no 6 pp 1161-78
[11] Thayer R E 1989 *The biopsychology of mood and arousal* (New York: Oxford University)
[12] https://developer.spotify.com/documentation/web-api
[13] Mcfee B, Raffel C and Liang D 2015 *Proc. of the Python in Science Conferences* (Austin: Texas) pp 18-25

IOP Conf. Series: Journal of Physics: Conf. Series **1195** (2019) 012010 doi:10.1088/1742-6596/1195/1/012010

Evolving and combining technical indicators to generate trading strategies

Chawwalit Faijareon and Ohm Sornil

School of Applied Statistics, National Institute of Development Administration, Bangkok, Thailand.

E-mail: chawwalit.f@nida.ac.th

Abstract. Technical analysis is a widely used approach for trading securities. Various indicators are used, such as moving average, stochastic oscillator and relative strength index. Applications of these indicators are typically based on experiences and rules of thumb which hardly are effective in general. This paper presents a technique for evolving indicator parameters using Non-Dominated Sorting Genetic Algorithm II and combining the indicators to generate a trading strategy. Experiments are conducted using actual stocks from the Stock Exchange of Thailand show that the proposed technique generates trading strategies that outperform other well-known techniques and is applicable to real world security trading.

1. Introduction

Traditional investment instruments include trading bonds, shares, futures, options, foreign exchanges, and precious metals. Among these securities, trading stocks has been the most popular alternative. Technical analysis focuses on price and volume movements of stocks [1]. Typically, traders use indicators, such as moving average, relative strength index, moving average convergence divergence and stochastic oscillator, to determine buy and sell signals. In addition, they may use chart patterns, such as price movement patterns and candlestick chart patterns, to predict future prices and trends.

It is challenging to achieve positive gains when the market involves a large number of investors trading against each other. The efficient market hypothesis confirms that advantages gained by an investor are vulnerable to be neutralized by others when they have access to the same kind of market information [1]. Investors then try to find extra information to help in trading and consider that historical data may provide indications of future price movements.

Trading rules derived from technical analysis have become the focus of many investors, especially high-frequency traders [2]. The rules have been mostly designed with parameters adopted from general traders' practices, such as the choices of durations in the moving average and relative strength index. Researches have begun to focus their attentions towards optimizing the trading rules. In [3-6], a genetic algorithm is employed to determine rules' coefficients, and in [7] a multi-objective optimization algorithm is employed to determine the coefficients. Moreover, it is investigated in [8] whether adaptive rules would make the investment more profitable.

This research presents a technique to generate strategies for trading stocks where Non-dominated sorting genetic algorithm II (NSGA-II) [9] is employed to determine the best parameters for technical indicators, and a decision tree is used to combine the indicators to create a strategy. The goal is to create a trading strategy that generates a high return on investment. The experiments are conducted on

actual stock price data from Stock Exchange of Thailand, and the results are compared with other popularly used strategies.

In the rest of the paper, six technical indicators are introduced; the proposed technique is presented; results of the experimental evaluations are shown and discussed; and finally conclusions are drawn.

2. Technical indicators used in the study

Six indicators are used in this research which consist of slope, exponential moving average (EMA), moving average convergent divergent (MACD), relative strength index (RSI), stochastic oscillators (STO) and average directional index (ADX).

2.1. Slope

Slope is a linear relationship between predictor variable and dependent variable. A slope can be positive or negative and calculated as:

$$m = \frac{(p_i - p_{i-n})}{(t_i - t_{i-n})} \tag{1}$$

where m is the slope, p_i is the price at time t_i. A sample trading rule based on slope is: buy if the n-day slope is positive. Parameter of this rule is n days.

2.2. Exponential moving average

Exponential moving average (EMA) is an average price in a specified period of time. The price of the last price changes and responds faster than simple moving average (SMA). It can be calculated as:

$$EMA_t(N) = \left[\frac{2}{n} \times \left(P_t - EMA_{t-1}(N)\right)\right] + EMA_{t-1}(N) \tag{2}$$

where EMA_t is EMA at time t, N is length of EMA, P is price at time t. A sample trading rule based on EMA is: buy if EMA N days is greater than price. Parameter of this EMA trading rule is N day.

2.3. Moving average convergent divergent

Moving average convergent divergent (MACD) was invented by Gerald Appel in late 1970's. It is used to track the direction of the stock and the force of the stock price by using two moving averages:

$$\begin{aligned} MACD &= EMA_{n_1} - EMA_{n_2} \\ MACD\ Signal &= EMA_{n_3}\ of\ MACD \end{aligned} \tag{3}$$

A sample trading rule based on MACD is: buy if MACD is greater than signal. Parameters of the MACD trading rules are fast length n_1, slow length n_2, and signal length n_3.

2.4. Relative strength index

Relative strength index (RSI) is an indicator used to track the direction of a stock and measure the rate of change in stock prices over a period of time. The value is between 0 and 100 and can be calculated as:

$$RSI = 100 - \frac{100}{1 + RS} \tag{4}$$

where RS is the average gain of up periods during a time frame divided by average loss of down periods during the time frame.

2.5. Stochastic oscillators

Stochastic oscillators (STO) are indicators used to analyze price movements over time. STO is usually used with short-term trading. The formula is calculated as:

IOP Conf. Series: Journal of Physics: Conf. Series **1195** (2019) 012010 doi:10.1088/1742-6596/1195/1/012010

$$\%K = \frac{Close_t - Lowest\ Low}{Highest\ High - Lowest\ Low} \times 100$$
$$\%D = n\ day\ SMA\ of\ \%K \tag{5}$$

where *Lowest Low* is the lowest low price in *n* day look back period, and *Highest High* is the highest high price in *n* day look back period. A sample trading rule based on STO is: buy if STO is greater than signal. Parameters of the STO trading rule are %K n_1, %D n_2.

2.6. Average directional index

Average directional index (ADX) is used to determine the direction or trend of the price and can be calculated as:

$$+DI = \frac{Moving\ Average\ of\ +DM}{True\ Range} \times 100$$
$$-DI = \frac{Moving\ Average\ of\ -DM}{True\ Range} \times 100 \tag{6}$$
$$ADX = Modify\ Moving\ Average\ of\ \left(\frac{(+DI)-(-DI)}{(+DI)+(-DI)} \right) \times 100$$

where *+DM* is the positive directional movement, *-DM* is the negative directional movement. A sample trading rule based on ADX is: buy if the positive directional index is greater than the negative directional index. Parameter of this trading rule is *n* days.

3. Proposed technique

In our proposed technique, for each indicator there are two trading rules associated with it. One is for buying, and the other is for selling. Parameters of each indicator are evolved by Non-dominated sorting genetic algorithm-II (NSGA-II) using 2 objective functions. Those rules are then combined by the Chi-square Automatic Interaction Detector (CHAID) algorithm to create a trading strategy. The total of 26 rules and their parameters are shown in Table 1.

Table 1. Trading rules

Indicators	Buy Rules	Parameters	Sell Rules	Parameters
Slope	Prince slope up in (n1) days	n1	Prince slope down in (n2) days	n2
EMA	Price more than EMA(n3)	n3	Price less than EMA(n4)	n4
EMA	EMA(n5) more than EMA(n6)	n5 , n6	EMA(n7) less than EMA(n8)	n7 , n8
EMA	EMA(n9) slope up in (n10) days	n9 , n10	EMA(n11) slope down in (n12) days	n11 , n12
MACD	MACD(n13,n14) more than Signal(n15)	n13 , n14 , n15	MACD(n16,n17) less than Signal(n18)	n16 , n17 , n18
MACD	MACD(n19,n20) more than Threshold(n21)	n19 , n20 , n21	MACD(n22, n23) less than Threshold(n24)	n22 , n23 , n24
MACD	MACD(n25,n26) slope up in (n27) days	n25 , n26 , n27	MACD(n28,n29) slope down in (n30) days	n28 , n29 , n30
RSI	RSI(n31) more than Signal(n32)	n31 , n32	RSI(n33) less than Signal(n34)	n33 , n34
RSI	RSI(n35) more than Threshold(n36)	n35 , n36	RSI(n37) less than Threshold(n38)	n37 , n38
STO	STO(n39) more than Signal(n40)	n39 , n40	STO(n41) less than Signal(n42)	n41 , n42
STO	STO(n43) more than Threshold(n44)	n43 , n44	STO(n45) less than Threshold(n46)	n45 , n46
ADX	ADX(n47) slope up in (n48) days	n47 , n48	ADX(n49) slope down in (n50) days	n49 , n50
ADX	+DI (n51) more than -DI(n51)	n51	+DI(n52) less than -DI(n52)	n52

4. Evolving trading rule parameters

In this research, the Non-Dominated Sorting Genetic Algorithm II (NSGA-II) algorithm is used to determine the optimal parameters for the rules. NSGA-II [6] is an optimization algorithm for finding

the most likely set of possible solutions to a problem while optimizing multiple objective functions simultaneously. It can be expressed as:

$$Minimize(or\ Maximize):\{f_1(x), f_2(x), ..., f_m(x)\} \tag{7}$$

where x is the vector of decision variables, $f_i(x)$ is a function of the objective i, NSGA-II returns a non-dominated set of answers, called a Pareto optimal set, where any answer x is better or not dominated by another answer y.

NSGA-II is an improved version of NSGA which can increase performance in the spread of solution and convergence near true-pareto optimal front. Deb K., Pratap A., Agarwal S. and Meyarivan T. [6] simulated several test problems from previous study using NSGA-II and claimed that this technique outperformed the two elitist MOEAs, i.e., PAES and SPEA. NSGA-II procedure can be described as follow:

- Create *N* initial, random populations.
- Calculate fitness values of each population.
- Rank the population by a non-dominated sorting.
- Calculate a Crowding Distance.
- Use a binary tournament selection, binary simulation crossover (SBX) and polynomial mutation for generating an offspring population.
- Combine parent population and offspring population.
- Rank the combined population and select *N* chromosomes by ranking and crowding distance for new generation.
- Check the terminating condition. If the condition is met, the last generation is assigned to the best set of solutions; else the procedure will continue by assigning the last population to the initial population.

The NSGA-II is used to determine the best parameters for trading rules. The example of chromosome encoding for MACD is shown in Table 2.

Table 2. Chromosome encoding

Indicator MACD					
Buy if MACD(*n13, n14*) more than Signal(*n15*)			**Sell if MACD(*n16, n17*) less than Signal(*n18*)**		
n day Fast (*n13*)	*n* day Slow (*n14*)	*n* day Signal (*n15*)	*n* day Fast (*n16*)	*n* day Slow (*n17*)	*n* day Signal (*n18*)

Maximum sensitivity and specificity are used as 2 fitness functions for NSGA-II in this research which can be expressed as:

$$\begin{aligned} Sensitivity_{Buy} &= \frac{TP_{Buy}}{TP_{Buy} + FN_{Buy}} \\ Specificity_{Buy} &= \frac{TN_{Buy}}{TN_{Buy} + FP_{Buy}} \\ Sensitivity_{Sell} &= \frac{TP_{Sell}}{TP_{Sell} + FN_{Sell}} \\ Specificity_{Sell} &= \frac{TN_{Sell}}{TN_{Sell} + FP_{Sell}} \\ Sensitivity &= Sensitivity_{Buy} \times Sensitivity_{Sell} \\ Specificity &= Specificity_{Buy} \times Specificity_{Sell} \end{aligned} \tag{8}$$

where TP_{Buy} is the number of Buy signals from the rule while the actual tomorrow price is up, TP_{Sell} is number of Sell signals from the rule while the actual tomorrow price is down, TN_{Buy} is number of Not Buy signals from the rule while the actual tomorrow price is down, TN_{Sell} is number of Not Sell signals

from the rule while the actual tomorrow price is Not Sell, FN_{Buy} is number of Not Buy signals from the rule while the actual tomorrow price is Not Buy, FN_{Sell} is number of Not Sell signals from the rule while the actual tomorrow price is down, FP_{Buy} is number of Buy signals from the rule while the actual tomorrow price is down, FP_{Sell} is number of Sell signals from the rule while the actual tomorrow price is up.

4.1. Creating trading strategy

The Chi-square Automatic Interaction Detector (CHAID) algorithm [10] is employed to create a trading strategy by combining rules generated from the previous step. CHAID is one of the decision tree algorithm for classification. CHAID is a multi-way tree which can split more than two nodes. The algorithm calculates independent chi-square tests to determine the p-value which is used to select the variable to separate the nodes. The statistics for the chi-square test of independence is

$$\chi^2 = \sum_{i,j} \frac{(O_{ij} - E_{ij})^2}{E_{ij}} \tag{9}$$

where i is 1, 2, ..., r and j is 1, 2, ..., c, O_{ij} is the observed(actual) frequency value, E_{ij} is the expected frequency value. The CHAID algorithm consists of three steps, i.e., merging, splitting and stopping as follow:

- Merging step will calculate a significant test on each categorical independent variable towards the dependent variables and merge most similar categories by selecting the smallest significant or biggest p-value of the category.
- Splitting step will separate a node by considering the biggest significant or smallest p-value of independent variables obtained from the merging process.
- Stopping step will checks the creation and separation of nodes, with the following requirements:
  - If the depth of tree reaches the maximum depth limit, the data classification process stops.
  - If the size of the node is less than the minimum node size, then the node will stop splitting.

5. Experimental evaluations

To evaluate the proposed method, data of 12 stocks from Stock Exchange of Thailand (SET) during 2008 to 2017 are used in the experiments. The data are divided into training data (from 2008 to 2015) and testing data (from 2016 to 2017). Amibroker software [11] is used for backtesting and measuring effectiveness of each strategy. The proposed method is compared with 8 popular trading strategies which are:

- Buy and Hold
- EMA: if EMA(5) crossover EMA(20) then buy. If EMA(20) crossover EMA(5) then sell.
- MACD(1): if Fast(12) crossover Slow(26) then buy. If Slow(26) crossover Fast(12) then sell.
- MACD(2): if MACD(12,26) crossover Signal(9) then buy. If Signal(9) crossover MACD(12,26) then sell.
- RSI: RSI(14) less than 30 then buy. If RSI(14) more than 70 then sell.
- STO(1): if Fast(14) crossover Slow(3) then buy. if Slow(3) crossover Fast(14) then sell.
- STO(2): if Slow(14,3) more than 20 then buy. if Slow(14,3) less than 80 then sell.
- ADX: if +DI crossover -DI then buy. If -DI crossover -DI then sell.

Table 3. Results of backtesting

STOCK	METHOD	Net Profit	Max. System Drawdown	STOCK	METHOD	Net Profit	Max. System Drawdown
ADVANC	Propose Method	**45.37%**	-19.93%	ITD	Propose Method	-19.49%	-29.59%
	Buy and Hold	33.57%	-25.26%		Buy and Hold	-44.62%	-50.71%
	EMA	20.68%	-22.86%		EMA	-19.41%	-25.94%
	MACD(1)	-4.88%	-25.04%		MACD(1)	-12.71%	**-20.23%**

	MACD(2)	28.14%	-23.73%		MACD(2)	-20.83%	-33.58%
	RSI	25.78%	-14.46%		RSI	-16.80%	-43.82%
	STO(1)	32.10%	**-11.79%**		STO(1)	**-4.65%**	-24.41%
	STO(2)	15.66%	-17.55%		STO(2)	-29.80%	-38.06%
	ADX	18.76%	-18.60%		ADX	-28.78%	-33.30%
AMARIN	Propose Method	2.29%	**-23.09%**	PTT	Propose Method	52.22%	-11.20%
	Buy and Hold	-16.36%	-46.00%		Buy and Hold	**84.87%**	-16.81%
	EMA	-14.15%	-30.14%		EMA	53.08%	-12.68%
	MACD(1)	-17.69%	-33.72%		MACD(1)	45.59%	-16.31%
	MACD(2)	-21.21%	-36.65%		MACD(2)	27.97%	-15.07%
	RSI	-8.38%	-36.29%		RSI	22.47%	**-8.40%**
	STO(1)	**10.96%**	-24.35%		STO(1)	26.22%	-13.61%
	STO(2)	-16.53%	-43.39%		STO(2)	16.41%	-9.17%
	ADX	-4.73%	-34.76%		ADX	65.19%	-12.24%
BBL	Propose Method	**54.87%**	-6.29%	SCC	Propose Method	8.57%	-21.56%
	Buy and Hold	34.67%	-15.79%		Buy and Hold	7.56%	-15.33%
	EMA	14.96%	-14.59%		EMA	-4.29%	-12.50%
	MACD(1)	25.57%	-14.58%		MACD(1)	-2.36%	-10.16%
	MACD(2)	-6.54%	-25.32%		MACD(2)	-9.24%	-23.77%
	RSI	7.19%	**-6.12%**		RSI	6.26%	**-8.20%**
	STO(1)	6.78%	-10.31%		STO(1)	2.09%	-17.57%
	STO(2)	4.49%	-12.68%		STO(2)	**22.56%**	-9.34%
	ADX	4.63%	-23.45%		ADX	-1.01%	-11.82%
BKI	Propose Method	**4.22%**	-6.62%	TDEX	Propose Method	**39.54%**	**-4.88%**
	Buy and Hold	0.85%	-7.40%		Buy and Hold	37.88%	-8.44%
	EMA	-4.73%	-7.89%		EMA	17.16%	-11.84%
	MACD(1)	-0.77%	-5.70%		MACD(1)	22.45%	-10.87%
	MACD(2)	-6.22%	-8.29%		MACD(2)	28.29%	-8.08%
	RSI	2.59%	**-3.43%**		RSI	16.71%	-6.25%
	STO(1)	-8.34%	-9.42%		STO(1)	30.69%	-6.45%
	STO(2)	-4.18%	-7.63%		STO(2)	6.75%	-6.45%
	ADX	-3.26%	-4.61%		ADX	21.56%	-8.04%
CK	Propose Method	1.69%	**-15.29%**	SCG	Propose Method	**81.30%**	-4.31%
	Buy and Hold	-5.36%	-25.55%		Buy and Hold	-3.20%	-13.27%
	EMA	-5.06%	-23.34%		EMA	-29.58%	-36.35%
	MACD(1)	-7.56%	-29.37%		MACD(1)	-15.63%	-19.91%
	MACD(2)	-13.63%	-29.66%		MACD(2)	-18.29%	-27.20%
	RSI	**13.86%**	-15.75%		RSI	0.00%	**0.00%**
	STO(1)	-15.05%	-25.53%		STO(1)	-4.30%	-28.00%
	STO(2)	-6.39%	-18.71%		STO(2)	-0.63%	-11.34%
	ADX	-2.01%	-24.23%		ADX	-14.35%	-18.52%
CPF	Propose Method	32.59%	**-15.26%**	VNG	Propose Method	**23.85%**	**-19.84%**
	Buy and Hold	**36.36%**	-29.02%		Buy and Hold	-20.29%	-40.11%
	EMA	-0.44%	-24.65%		EMA	-5.96%	-26.82%
	MACD(1)	11.33%	-29.80%		MACD(1)	-9.92%	-28.82%
	MACD(2)	-1.41%	-25.69%		MACD(2)	2.72%	-26.89%
	RSI	-4.50%	-21.98%		RSI	22.53%	-17.04%
	STO(1)	0.72%	-26.91%		STO(1)	-3.82%	-32.35%
	STO(2)	-7.29%	-23.54%		STO(2)	-8.61%	-30.16%
	ADX	-1.54%	-23.65%		ADX	-7.01%	-28.10%

The results of testing with data of 12 stocks for 2 most recent years (2016 to 2017) are shown in Table 3. Net Profit is the net profit achieved after completing all trades in the period, and Max System Drawdown is the largest peak to valley percentage decline experienced while trading in the period. We can see that strategies generated from the proposed technique yield the highest returns or are ranked among the top strategies for stocks that are in an upward trend. Also the generated strategies suffer the lowest losses or are among the top strategies for stocks that are in a downward trend. For some stocks, there may be a very few strategies that yield higher returns than our strategies; but there is no common strategy that generally outperforms our technique.

Overall, strategies generated from the proposed technique outperform 8 other comparative techniques which mean that our technique is applicable to trading actual stocks.

IOP Conf. Series: Journal of Physics: Conf. Series **1195** (2019) 012010 doi:10.1088/1742-6596/1195/1/012010

6. Conclusion

This paper presents a technique to generate effective trading strategies using a combination of a multi-objective genetic algorithm and a decision tree. A strategy uses rules from 6 popularly used technical indicators which are slope, exponential moving average, moving average convergence divergence, relative strength index, stochastic oscillator and directional index. Appropriate parameters for each rule are determined by NSGA-II with sensitivity and specificity used as two objective functions. A trading strategy is then created by combining the evolved rules together via the CHAID algorithm. The technique is compared with eight popularly used strategies on 12 random stocks from Stock Exchange of Thailand. The results show that the proposed technique generates trading strategies that outperform other strategies, and the technique is applicable to real-world security trading.

References

[1] Lee C I, Pan M S and Liu Y A 2001 On market efficiency of Asian foreign exchange rates: evidence from a joint variance ratio test and technical trading rules *Journal of International Financial Markets, Institutions and Money* **11** 199-214.

[2] Fyfe C, Marney J P and Tarbert H F E 1999 Technical analysis versus market efficiency - a genetic programming approach *Applied Financial Economics* **9** 183–191.

[3] Chen J S 2005 Trading Strategy Generation Using Genetic Algorithms *Asian Journal of Information Technology* **4(4)** 310-322.

[4] Kwok N M, Fang G and Ha Q P 2009 Moving Average-Based Stock Trading Rules from Particle Swarm Optimization *2009 International Conference on Artificial Intelligence and Computational Intelligence.*

[5] Kraithong R and Sornil O 2014 A Cooperative Coevolution Genetic Algorithm for Generating Security Trading Strategies *International Journal of Applied Engineering Research* **9** 17859-17869.

[6] Simões A, Neves R F and Horta N 2010 An Innovative GA Optimized Investment Strategy based on a New Technical Indicator using Multiple MAS *International Conference on Evolutionary Computation.*

[7] Fayek M B, El-Boghdadi H M and Omran S M 2013 Multi-objective Optimization of Technical Stock Market Indicators using GAs *International Journal of Computer Applications* **68** 41-48.

[8] Ellis C A and Parbery S A 2005 Is smarter better? a comparison of adaptive, and simple moving average trading strategies *Research in International Business and Finance* **19** 399-411.

[9] Deb K, Pratap A, Agarwal S and Meyarivan T 2002 A fast and elitist multiobjective genetic algorithm:NSGA-II *IEEE Transactions on Evolutionary Computation* **6** 182-197.

[10] Kass G V 1980 An Exploratory Technique for Investigating Large Quantities of Categorical Data *Applied Statistics* **29** 119-127.

[11] Bandy H B 2012 *Introduction to AmiBroker (2nd ed.)* South Dakota (Blue Owl Press, Inc.)

IOP Conf. Series: Journal of Physics: Conf. Series **1195** (2019) 012011 doi:10.1088/1742-6596/1195/1/012011

A hierarchical system for optimising a dynamic system of traffic crossroads control using an expert system

Jakub Gaj[1], Bogdan Walek[1]*, Radim Farana[2] and Martin Kotyrba[1]

[1] Department of Informatics and Computers, University of Ostrava, Ostrava, Czech Republic
[2] Department of Informatics, Mendel University in Brno, Brno, Czech Republic

*bogdan.walek@osu.cz

Abstract. Nowadays in the field of the automotive industry and urban public transport, there are a lot of challenges and problems to solve. Every day there are closures, roadworks, and other aspects of public transport, which cause traffic jams. This paper deals with a hierarchical system for optimising a dynamic system of traffic crossroads control using an expert system. The hierarchical system uses an expert system constructed in a few layers which evaluate the length of the green signal for each phase of the crossroad. The main goal of the proposed system is traffic optimisation at crossroads and reduction of traffic jams. The proposed system was verified on a real crossroad in the city of Ostrava.

1. Introduction
Nowadays in the field of the automotive industry and urban public transport, there are a lot of challenges and problems to solve. Every day there are closures, roadworks, and other aspects of public transport, which cause traffic jams. People spend many hours a year in these jams. That is why every urban transport department tries to optimise traffic junctions, crossroads, and overall traffic to make the jams as small as possible. There is a number of approaches that attempt scientifically different ways of optimising and managing traffic flows [1-4].

Currently, the crossroads control is usually set dynamically [5-7]. This means that we have created a signalling plan for each phase of a given junction and it switches to the basis of the evaluation of the traffic flow of vehicles, which approach the junction in each direction. Accordingly, the signal marking of traffic lights is switched.

In the event of a failure in the dynamic system, the traffic light control is switched over to the standby static signalling plan, in which the switching of green light is fixed for each phase [8].

The last option for traffic control is the traffic policeman at the crossroads, managing the situation according to the subjective point of view and representing the role of an expert.

However, traffic jams are still the main problem of the transport control. Thus, it will be suitable to supplement these systems with artificial intelligence to replace the expert's position and dynamically manage the transport. This paper focuses on dynamic management of crossroads through a hierarchical system (using an expert system) traffic optimisation at crossroads and reduction of traffic jams.

2. Methodology

IOP Conf. Series: Journal of Physics: Conf. Series **1195** (2019) 012011 doi:10.1088/1742-6596/1195/1/012011

The structure of the proposed hierarchical system is visually shown in Figure 1. The hierarchical system uses a set of expert systems for evaluating the importance of specific phases in the crossroad.

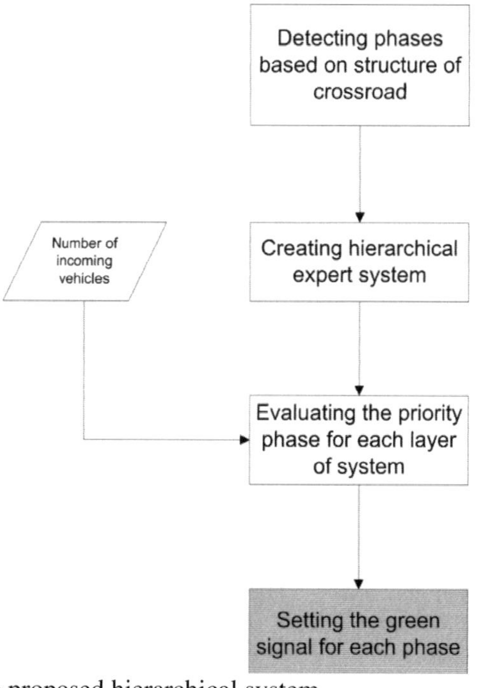

Figure 1. The structure of the proposed hierarchical system

Main steps of the proposed system are more described in the subsequent subsections.

2.1. *Detecting phases based on the structure of the crossroad*

In the first step, the traffic streams and their directions at the crossroad will be detected. Based on detected directions, the phases are inputs to the hierarchical expert system.
An example of a crossroad is shown in Figure 2.

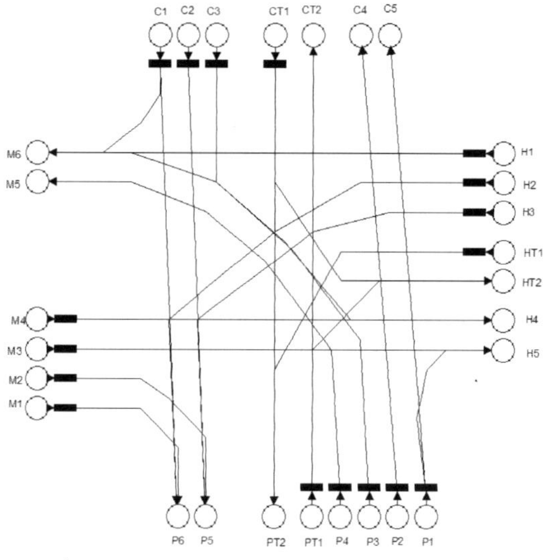

Figure 2. Example of a crossroad

There are following directions detected at the crossroad. From P1 we can turn to H5 or go straight to C5, from P2 we can go straight to C4, from P3 we can turn to M6, from P4 we can turn to M5. PT1 is a tram and a tram can turn to HT2 or go straight to CT2, from M1, we can turn to P6, from M2 we can turn to P5, from M3 we can go straight to H5, from M4 we can go straight to H4, from C1 we can turn to M6 or go straight to P6, from C2 we can go straight to P5, from C3 we can turn to H4. CT1 is a tram and a tram can go straight to PT2 or turn to HT2, from H1 we can go straight to M6, from H2 we can turn to P6, from H3 we can turn to P5. HT1 is a tram and a tram can only go to PT2.

In Figure 3, there is a satellite map of the sample crossroad schematically shown in Figure 2 and shows the various driving possibilities.

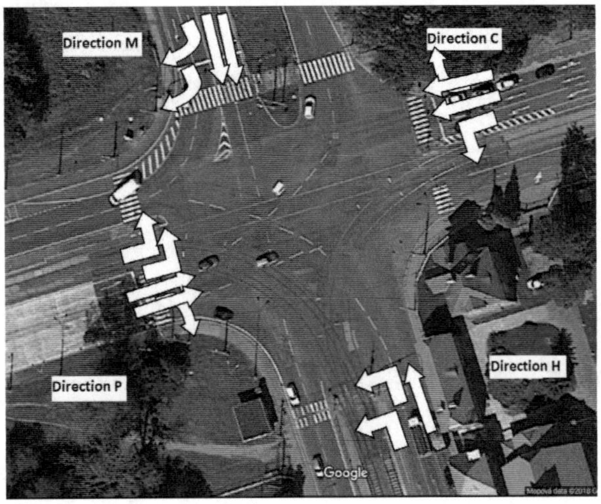

Figure 3. Satellite map of the sample crossroad

2.2. *Creating a hierarchical expert system*

In this step, the phases detected in the previous steps will be used as input variables to the hierarchical expert system. Based on the number of detected phases, the specific layers of the hierarchical expert system will be created. For the first layer, the input variables are all detected phases. For other layers, the input variables are all the remaining phases after evaluating the previous layer.

For each phase, the workload of the phase will be recognised. The workload of the phase will be captured by traffic sensors that are along the roadway in each direction. In each direction, the sensors are in two waves to provide correction data. For our purposes, we have set the workload as "high", "medium" and "low".

The input data for each variable will be the number of cars at that phase. The individual phase is the start of single traffic lights for selected lanes.

Each input variable will have a range for a single cycle. The minimum value will be set to 0 cars, but maximum values are set to 10 - 20 - 40 cars per cycle. Thus, the mean value will be between 5 and 10-20 cars, see Figure 4.

For creating a knowledge base of the expert system, the LFLC tool can be used. This tool is able to define input and output linguistic variables and IF-THEN rules. LFLC tool also has inference mechanisms and implemented defuzzification procedures, so we can create a complete expert system with this tool. The LFLC tool is more described in [9].

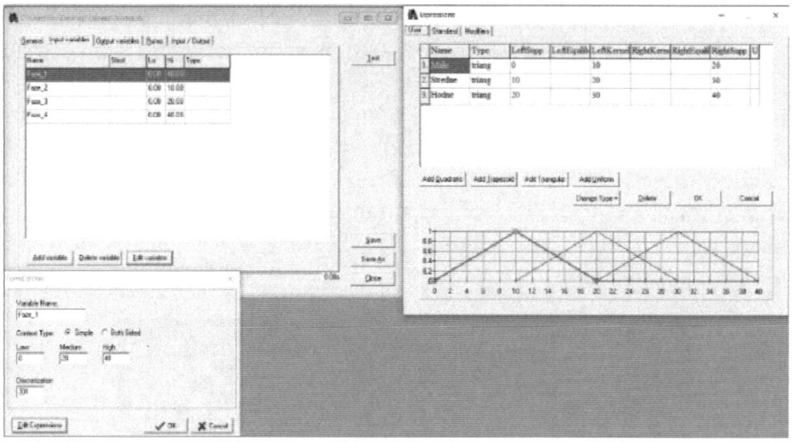

Figure 4. Input variable Phase 1 of the proposed expert system

2.3. Evaluating the priority phase for each layer of the system

Depending on the use of the individual phases, we will define the strategy to determine the priority phase that will have the majority part of the overall crossroads cycle.

Then, based on the number of phases the strategies are defined. Here are strategies for four phases:

- Strategy S0 - in this case, the cycle will remain in the standard way
- Strategy S1 - the majority part will be set at Phase 1
- Strategy S2 - the majority part will be set at Phase 2
- Strategy S3 - the majority part will be set in Phase 3
- Strategy S4 - the majority part will be set at Phase 4

Examples of IF-THEN rules for the first layer are:

```
IF (PHASE 1 IS HIGH) AND (PHASE 2 IS HIGH) AND (PHASE 3 IS HIGH) AND
(PHASE 4 IS HIGH) THEN STRATEGY IS S0
IF (PHASE 1 IS HIGH) AND (PHASE 2 IS HIGH) AND (PHASE 3 IS LOW) AND
(PHASE 4 IS LOW) THEN STRATEGY IS S1
IF (PHASE 1 IS LOW) AND (PHASE 2 IS HIGH) AND (PHASE 3 IS HIGH) AND
(PHASE 2 IS LOW) THEN STRATEGY IS S3
IF (PHASE 1 IS LOW) AND (PHASE 2 IS LOW) AND (PHASE 3 IS HIGH) AND
(PHASE 4 IS HIGH) THEN STRATEGY IS S4
```

2.4. Hierarchical evaluating by the expert system

Evaluation by the expert system is provided hierarchically. For each layer, the output variable is the strategy which defines the priority phase (with a major part of the overall crossroads cycle). For the second and next layer, the input variables are all remaining phases, and the output variable is again strategy.

For instance, if there are four phases at the crossroad, the evaluating will be done as follows:

1. There are four phases: Phase 1, Phase 2, Phase 3, Phase 4 and three layers
2. For the first layer there are four input variables, and after evaluation, the priority phase is set by the evaluated strategy (for example, strategy S4 represents Phase 4)
3. For the second layer, there are three input variables representing all phases except the priority phase of the previous layer evaluation (Phase 1, Phase 2, Phase 3) and after evaluation, the priority phase is set by the evaluated strategy (for example, strategy S3 represents Phase 3)
4. For the third layer, there are two input variables represent all phases except the priority phase of the previous layer evaluation (Phase 1, Phase 2) and after evaluation, the priority phase is set by the evaluated strategy (for example, strategy S2 represents Phase 2)

The process of evaluation is more described below.

In the first layer of our system, the expert system will evaluate the traffic situation, which is the most difficult. In the next step, the expert system selects the stage of the signalling plan which is to be activated to ease the traffic situation. After selecting the phase called “priority phase", we set the largest time period of the entire cycle of a specific crossroad, namely 60% of the real cycle. The length of the real cycle, in which the signal is green in each direction, is determined by the saturation flow method. The second layer will again perform the same procedure as in the first layer but without the first selected phase. After selecting the second priority phase, we set the length of the signal so that it gets again 60% of the time, but now from the remaining time after the first layer. Then the system switches to the third layer. In this layer, we only select from two phases that will be a priority. This phase then reaches 60% of the remaining time after the second layer and the remaining time gets the phase only the least busy. We have to ensure that each phase in a single cycle is done and there is no situation that in some direction the red signal will be still on. Therefore, we will observe that each phase must be at least 5% of the total cycle time. The output of the expert system will be a single time setting for the individual signal markings of the test crossroad.

2.5. Setting the green signal for each phase

After evaluation of all layers of the hierarchical expert system, the length of the green signal for each phase will be defined.

3. Verification

Verification of the expert system was carried out on data obtained from the city of Ostrava, which measured the traffic through the crossroad shown in Figure 3.
The measurements took place in four cycles at the rush hour, from 15:00 to 16:00.
The evaluation by the hierarchical expert system is shown in Figure 5.

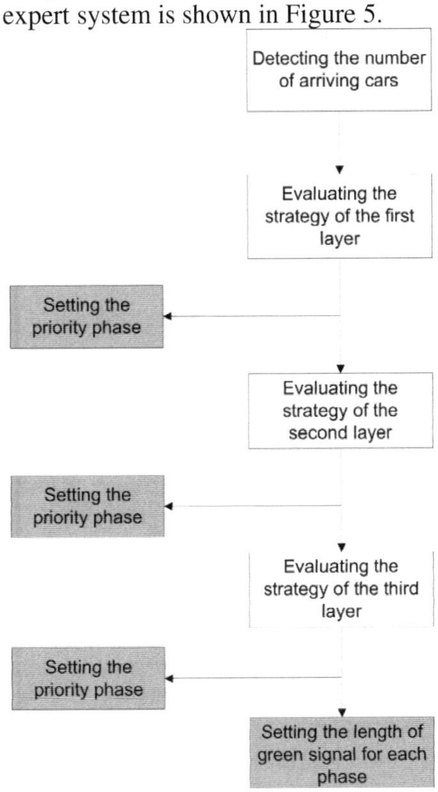

Figure 5. Evaluation by the hierarchical expert system

Table 1 shows the numbers of cars in each phase of the given cycles.

IOP Conf. Series: Journal of Physics: Conf. Series **1195** (2019) 012011 doi:10.1088/1742-6596/1195/1/012011

Table 1. Number of cars in each phase

Cycle	Phase 1	Phase 2	Phase 3	Phase 4
C1	22.7	5.8	11.6	33
C2	24.9	6.3	15.7	34.1
C3	34.5	4.1	5.2	19.3

The outputs of the first expert system layer for each cycle are shown in Table 2.

Table 2. The output of the first expert system layer

Cycle	Strategy
C1	S4
C2	S0
C3	S1

From the table, we can conclude that for cycles C1 and C3 the strategy is chosen to divide the time for one of the phases in the majority of ways. For C2, strategy S0 is chosen, which means that the time division for each phase will be done in a standard way.
For cycles C1 and C3 and their resulting S4 and S1 strategies, this method will be used to calculate the times for the individual phases (analogously would be used for strategies S2 and S3):

- For the phase that represents the output of the first layer of the expert system, 60% of the total time determined on the total intersection cycle
- All cycles outside the cycle representing the first phase output will be used to enter the second layer
- For the phase that represents the output of the second layer of the expert system, 60% of the remaining time will be used
- The remaining cycles will be used to enter the third layer
- For the phase that represents the output of the third layer of the expert system, 60% of the remaining time will be used

The following tables show a multilayer calculation for cycle C1. After selecting the resulting strategy of the first layer shown in Table II, all phases will be input into the calculation of the second layer except for the phase representing the first phase output. The second layer input variables and the resulting strategy of the second layer of the expert system are shown in Table 3.

Table 3. The output of the second expert system layer for cycle C1

Cycle	Phase 1	Phase 2	Phase 3	Strategy
C1	22.7	5.8	11.6	S3

Thus, the table shows that for phase 3, 60% of the time remaining after reading the first layer output was used (strategy S3 was chosen).

After selecting the resulting second layer strategy, the remaining two phases - Phase 1 and Phase 2 - will enter the third layer. The third layer input variables and the resulting strategy of the third layer of the expert system are shown in Table 4.

Table 4. The output of the third expert system layer for cycle C1

Cycle	Phase 1	Phase 2	Strategy
C1	22.7	5.8	S1

Thus, the table shows that for phase 1, 60% of the time remaining after calculating the output of the first and second layers will be used.

Calculation for cycle C3 will also be performed. The second layer input variables and the resulting strategy of the second layer of the expert system are shown in Table 5.

Table 5. The output of the second expert system layer for cycle C3

Cycle	Phase 1	Phase 2	Phase 3	Strategy
C3	4.1	5.2	19.3	S3

Thus, the table shows that for phase 3, 60% of the time remaining after calculating the first layer output was used (strategy S3 was chosen).

After selecting the resulting strategy of the second layer, the remaining two phases - Phase 1 and Phase 2 - will enter the third layer. The third layer input variables and the resulting strategy of the third layer of the expert system are shown in Table 6.

Table 6. The output of the third expert system layer for cycle C3

Cycle	Phase 1	Phase 2	Strategy
C3	5.2	4.1	S0

4. Conclusion

This paper deals with the area of crossroads traffic control.

In the introduction, the current state of crossroads traffic control was outlined. Then, a hierarchical system for optimising the dynamic system of traffic crossroads control was proposed. The hierarchical system uses an expert system constructed in few layers which evaluate the length of the green signal for each phase of the crossroad. The proposed system was described and the main parts of the system were explained.

The proposed system was verified on a real crossroad in the city of Ostrava.

In future work, our attention will be drawn to the verification of the proposed system at other crossroads.

References

[1] K. Han, H. Liu, V. V. Gayah, T. L. Friesz, T. Yao, "A robust optimization approach for dynamic traffic signal control with emission considerations", Transportation Research Part C: Emerging Technologies, 2016, 70, pp. 3-26.

[2] Bi. Y. Srinivasan, D. Lu, X. Z. Sun, W. Zeng, "Type-2 fuzzy multi-intersection traffic signal control with differential evolution optimization", Expert systems with applications, 2014, Volume 41, Issue 16, pp. 7338-7349.

[3] W. Turek, L. Siwik, M. Kisiel-Dorohinicki, S. Łakomy, P. Kala, A. Byrski, "Real-time metaheuristic-based urban crossroad management with multi-variant planning", Journal of Computational Science, 2017, 23, pp. 240-248.
[4] N. R. Yusupbekov, A. R. Marakhimov, H. Z. Igamberdiev, S. X. Umarov, "An Adaptive Fuzzy-Logic Traffic Control System in Conditions of Saturated Transport Stream", The Scientific World Journal, 2016.
[5] M. Papageorgiou, C. Diakaki, V. Dinopoulou, A. Kotsialos, Y. Wang, "Review of road traffic control strategies", Proceedings of the IEEE, 2003, pp. 2043-2067.
[6] A. Karoonsoontawong, S. T. Waller, "Integrated network capacity expansion and traffic signal optimization problem: robust bi-level dynamic formulation", Networks and Spatial Economics, 2010, Volume 10, Issue 4, pp. 525-550.
[7] F. F. Pashchenko, A. V. Kamenev, D. S. Kholodilov, "Modeling and control of traffic flows", Procedia Computer Science, 2015, 63, pp. 89-95.
[8] J. Gaj, M. Kotyrba, E. Volna, "Possibilities of Control and Optimization of Traffic at Crossroads Using Petri Nets", In Information Science and Applications (ICISA) 2016, Singapore, 2016, pp. 21-29.
[9] H. Habiballa, V. Novák, A. Dvořák, V. Pavliska, "Using software package LFLC 2000", 2nd International Conference Aplimat 2003, Bratislava, 2003, pp. 355-358.

Acknowledgments
This work was supported within the framework of the IT4Innovations Centre of Excellence project, reg. no. CZ.1.05/1.1.00/02.0070 supported by Operational Programme "Research and Development for Innovations" funded by Structural Funds of the European Union and state budget of the Czech Republic. The work was created during the completion of a Student Grant SGS02/UVAFM/2018 with student participation, supported by the Czech Ministry of Education, Youth and Sports.

IOP Conf. Series: Journal of Physics: Conf. Series **1195** (2019) 012012 doi:10.1088/1742-6596/1195/1/012012

A mission oriented reconfiguration technology for Spaceborne FPGA

J X Qin, J Yang*, Z Qu and Y X Wang

College of Artificial Intelligence at National University of Defense Technology, Changsha, Hunan, China.

Email: iohn323@163.com

Abstract. Reconfigurable technology based on FPGA is increasingly used in satellites to meet the demand of the increasing missions that satellites need to complete in recent years. In this paper, a mission oriented reconfiguration technology for spaceborne FPGA is introduced to address the problem that how to meet the demand of satellite reconfiguration mission to orbit, execution time and FPGA resource requirements. Satellite reconfiguration system is divided into three parts, that is, mission management module, reconfiguration management module and reconfigurable FPGA. Firstly, a mission list is set up to record the mission's execution time and space attributes, hardware space attributes and priority attributes. The mission is allocated to the FPGA resource area correspondingly based on the flexible region model. According to the time attribute and priority attribute of the mission, the reconfiguration management module carries out the reconfiguration action. And the reconfiguration management module mainly deals with reconfiguration library management and the entire or partial reconfiguration according to the configuration file. After the reconfigurable bit files are loaded to FPGA, the transformation and execution of the hardware resources from the mission to the hardware are completed. Finally, an evaluation of the real-time performance, reliability and security of the design is carried out. Experiments that based on a verification system composed of raspberry/PI and ZYNQ are conducted and results fully prove that the design in this paper can support mission oriented spaceborne FPGA reconfiguration.

1. Introduction

The payload of traditional satellite is solidified with a poor flexibility. For the reason that the satellite can only play a role in the specific orbit, and the other time, which takes up the most part, is in the closed state, resulting in an idle and waste of the satellite resources. The application of reconfigurable technology to satellites is a good way to solve this problem. While the reconfigurable technology in the ground is more mature, it has only started in recent years for satellites. A reconfigurable Satellite TT&C (telemetry, track and command, TT&C) communication system based on FPGA is built in [1]. Its goal is to realize the real-time update of the small satellite communication load function and the switch of communication mode through the configuration of the ground communication station. The software defined satellite "quantum" project, conducted by the ESA (European Space Agency, ESA), the European Communication Satellite Corporation and the Airbus defense and Aerospace Corp, which is expected to be launched in 2019, use the FPGA to reconfigure the function of the payload as in [2]. However, the reconfiguration of the satellite is based on the specific mission requirements. The FPGA resource requirements of the mission, the track and time of the mission execution are different, and the priority is also different. How to map missions to specific hardware resources, and complete

reconfiguration according to the needs and characteristics of satellites, there are few papers relatively on this topic. Nevertheless, the reconfiguration technology on the ground is more mature. There are many researches on mission placement. For example, a reconfigurable hybrid mission scheduling algorithm is introduced to reduce configuration time in [3]. Reference [4] discusses several reconfigurable resource management models, and proposes a one dimensional online scheduling model, which focuses on mission scheduling. In order to solve reconfigurable computing, mission scheduling system based on spatio temporal context is studied in [5]. Considering the actual situation of the satellites, the mission oriented FPGA reconfiguration technology is studied in this paper.

The missions of satellite will be more and more. Using reconfiguration technology to execute multiple missions can effectively save resources and improve resource utilization. Satellite reconfigurable systems are mainly divided into two parts: satellite to ground reconfiguration loop and satellite reconfiguration system [6]. This paper only considers satellite reconfiguration system, and mainly solves the problem that how to map the satellite mission to the hardware resources, and how to execute multiple reconfiguration missions. The work based on this paper can not only support the reconfiguration of the satellite borne FPGA, but also enable the satellite to gain a kind of certain autonomy, and can satisfy some special users like in the overseas and the public seas that are inconvenience of controlling the satellites, and even support the sudden reconfiguration of the real-time updating application scene.

2. Mission oriented FPGA reconfiguration system

FPGA reconfiguration system, in this paper, specifically refers to modules containing FPGA resources, not a specific load or single chip. However, in order to facilitate the experiment, this paper will consider the selection of FPGA, which will be explained later. Mission oriented refers to reconfiguration activities centered on missions and aims at achieving the goals of missions, rather than reconstructing them solely for reconfiguration. The mission of the satellite should be dynamic and diverse. For specific track position, in specific time period and specific application scene, the mission is different, and the reconfiguration also has the corresponding change. For example, when the satellite moves over the land and sea, the remote sensing algorithm is different; the detection algorithm is different from the different seasons over the same land.

In this paper, the spaceborne FPGA reconfiguration system is divided into three parts, that is, on-board computer mission management module, reconfiguration management module and reconfigurable FPGA. As in figure 1, a list of missions is set up for satellites, recording the hardware space properties of each mission, executing space-time attributes and priority attributes. According to the hardware space attribute, the mission is placed to the hardware resources based on the region model. The mission reconfiguration time is determined according to the execution time attribute and priority attribute. The reconfiguration management module manages reconfigurable bit files, and the reconfigurable bit files are loaded into reconfigurable FPGA to complete the reconfiguration mission according to the configuration file.

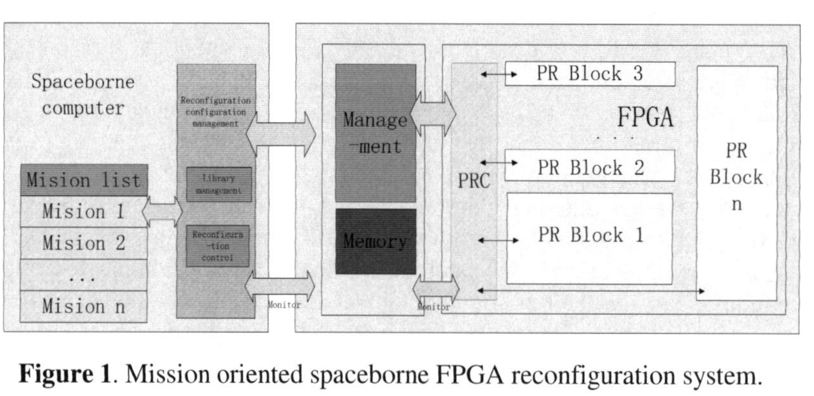

Figure 1. Mission oriented spaceborne FPGA reconfiguration system.

2.1. Mission management

Satellite On-board computer mission management module is responsible for managing satellite reconfiguration missions. The mission referred in this paper specifically refers to the mission of reconfigurable FPGA, because not all missions need FPGA for reconfiguration. A mission list is set up for the reconfigurable missions carried by the satellite, and the hardware space attributes, space-time attributes and priority attributes are registered. According to the hardware space properties of reconfiguration missions, the first step is to put reconfiguration missions into specific FPGA resource blocks. As shown in figure 2, there are three reconfigurable resource model diagrams, followed by one-dimensional model map, as in (a), two-dimensional model map, as in (b) and regional model map [4] as in (c). The peripheral bounding box represents the FPGA resources that can be reconfigured, and the small rectangles represent the FPGA resources occupied by the mission. The one-dimensional model is first proposed by [7], which simplifies placement and scheduling, and is suitable for Xilinx Virtex implementation. However, this model will lead to fragmentation inside and outside the FPGA, and need to adopt defragmentation technology. It is first proposed in [8] that the reconfigurable surface is divided into a static and fixed number block in a two-dimensional model. Each block can hold one mission at a time. This area partition simplifies placement and scheduling, and is conducive to online addressing. But its drawback is internal fragmentation. When a mission requires less than one block of resources, other resources in the block are wasted.

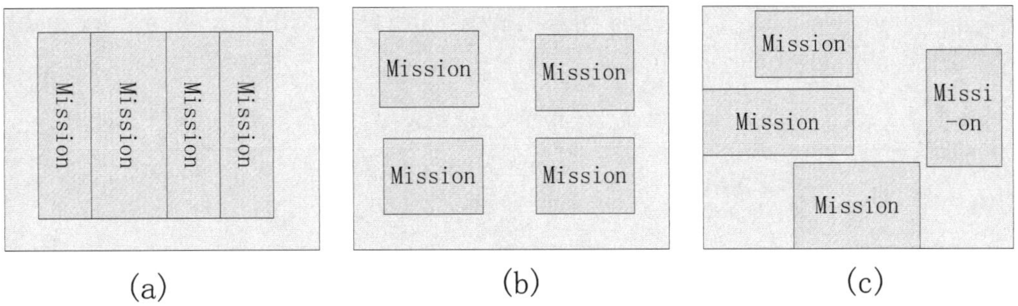

Figure 2. Reconfigurable resource model diagram.

In this paper, region model diagram is used based on the method proposed by Busquere J P et al in [9]. This method is the most flexible and can be placed in any size resource for the size of the mission requirement resource, so the resource utilization rate is high. It only needs the reservation space as the communication channel. According to the resource requirements of specific reconfiguration missions, specific areas can be delineated, and then loaded and run in real time. Moreover, when we need to update the overall layout, we can use entire reconfiguration to achieve overall resource reconfiguration. But entire reconfiguration needs to interrupt part of the program's running, so it is good advice to try to use partial reconfiguration.

Reconstructing the temporal and spatial attributes of a reconfiguration mission is the question of where the mission's executed orbit is and whether it is executed after it reaches the area. Considering the two attributes of reconfiguration mission, we can sort the mission and control and execute the reconfiguration mission according to sorting. Mission management divides missions into current missions, planning missions and appointment missions. The current mission is a running mission in FPGA. There can be one or more missions, with multiple current missions because FPGA can be divided into multiple reconfiguration areas and support multiple current mission parallel processing. Planning missions are missions that are sorted in the mission list in order of execution. The appointment mission is the mission to be reconstructed and implemented next. It also allows multiple appointment missions under the support of FPGA resources.

2.2. Reconfiguration management

Reconfiguration management module is mainly used for reconfigurable bit file library management and reconfiguration control. In the ground system, FPGA reconfiguration in the system can be easily realized by replacing the program storage chip directly or using the erasable memory for off-line/online programming. But in satellite, the reconfiguration of FPGA program needs some special restrictions, such as the non disassembly of the equipment after launching, the memory chip to resist space irradiation, the limited communication channel resources from the ground to the satellite. To store reconfigurable bit file libraries and to ensure the correctness and reliability of reconfigurable bit files, configuration data and configuration process must be a key concern for satellites.

The reconfigurable bit file can be stored in advance or later uploaded by the ground control station. All reconfigurable bit files are stored in the highly reliable memory of the on-board computers to resist space irradiation. In order to ensure the reliability of reconfiguration, satellites should avoid the possibility that executing reconfiguration while receiving the reconfiguration bit file. As shown in table 1, the scale of FPGA commonly used in aerospace is 3 million gates [10]. Calculated according to the commonly used uplink remote code rate 4000bit/s, the transmission time takes 2900 seconds. Such a long time is very prone to errors in the process of reconfiguration. Therefore, we need to upload first and then reconfigure. When the reconfiguration is started, this paper adopts the way that the spaceborne computer is transmitted to DDR through the network port, and then is written directly to FPGA through ARM control. Because the reconfiguration control of FPGA requires the coordination of ARM. The on-board computer is only responsible for the management of reconfigurable bit files and configuration files. The actual control of FPGA file writing and boot startup requires ARM control.

Table 1. Configuration data size and transmission time of FPGA used in spacecraft [10].

Type of FPGA	Scale/ten thousand gates	Amount of the configuring data/bit	Storage space/Mbit	Transmission time/s
XQVR300	30	1751808	16	500
XQVR1000	100	6127744	16	1500
XQR2V3000	300	10494368	16	2900
XQR2V6000	600	21849568	32	5900
XQR4VSX55	550	22744832	32	6000
XQ5VFX130T	1300	49234944	49	13000

In this paper, we verify the integrity and correctness of reconfigurable bit files. Table 2 is the structural diagram of the reconfigurable bit file, and table 3 is the first structure diagram of partial reconfiguration. As can be seen from the structure, the reconfiguration bit file can have one or more partitions corresponding to the partial reconfiguration. The Header Authentication is a validation of the header and can be selected. In each Partition, Checksum is used to check the files. After uploading a file, it is necessary to check whether it meets the required header and tail before loading it. The header is different or incorrect, indicating errors in the transmission and storage process. The tail is different or incorrect, maybe because there is no transmission to complete, or other problems occur. If the check is not passed, reconfiguration is not allowed.

Table 2. Structure of BOOT.bin [11].

Boot Image Header	Register Initialization	Image Header Table	Partition Header Table	Header Authentication (Optional)	Encrypted FSBL	FSBL RSA authentication (Optional)	Partition	…

Table 3. Structure of Partition Header [11].

Name	Partition Data Word Length	Extracted Data Word Length	Total Partition Word Length	Destination Load Address	Destination Execution Address	Data Word Offset in Image	Attribute Bits	Section Count	Checksum Word Offset	…
offset	0x0	0x4	0x8	0x0C	0x10	0x14	0x18	0x1C	0x20	…

Reconfiguration management is responsible for executing the reconfiguration action to the reconfigurable hardware. For FPGA to complete a reconfiguration, we need to consider its reconfiguration type, file loading, boot and reset power on. The reconfiguration types are divided into entire reconfiguration and partial reconfiguration. Entire reconfiguration is a redistribution of the whole resources of FPGA, which will affect the current running program. Partial reconfiguration can achieve partial reconfiguration without affecting the normal operation of other functions. The boot and reset power on are designed for specific hardware models and requirements. The key to reconfigure FPGA's logical resources is to reorganize its combinational logic. In the FPGA design of reconfigurable hardware units, the PROM/FLASH is configured to store FPGA software and hardware design files, and then by receiving or prestoring the design files, after receiving the reconfigured instructions, the reset signal is sent by the configuration management module, the FPGA power is restarted and the bits are reloaded into the FPGA, and the logical resources are reloaded. All refurbishing is carried out to achieve entire reconfiguration. The partial reconfiguration does not require power break. By addressing, the bit files that were reconfigured are loaded into the FPGA by ARM control to realize the reorganization of the logical resources in this part of the region.

In order to achieve partial reconfiguration of the function after confirming the FPGA resource area of the mission, we need to design the interface at the same time. This paper is based on ZYNQ, and the structure of partial reconfiguration is shown in figure 3. A Reconfigurable Partition (RP) is opened up in the FPGA resource of ZYNQ, and the RP is an instantiation of reconfigurable functional parts, and the Reconfigurable Module (RM) is distributed in the region. A RP can have one or more RM in it. When the RP is reconfigurable through the AXI4-Lite bus and the control instruction is sent, the other programs outside can continue to run in parallel and will not be interrupted by this part of the reconfiguration. The control of this part is completed by the configuration management module of ARM and the on-board computer.

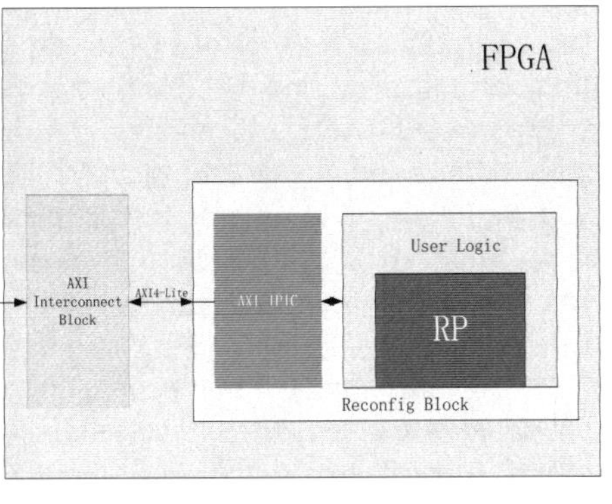

Figure 3. Architecture of partial reconfiguration [12].

Table 4. Mission list.

Mission list	Distri -ct	Priority	Place
ADSB	District 1	1	Place A
Add	District 2	2	Place B1
Product	District 3	2	Place B1
……	……	… …	……

3. Experimental verification

In this paper, an entire reconfiguration scenario and a partial reconfiguration scenario are set up to simulate two missions to verify the scheme. The first mission is ADSB, and the other mission, Add

IOP Conf. Series: Journal of Physics: Conf. Series **1195** (2019) 012012 doi:10.1088/1742-6596/1195/1/012012

and Product represents the signal process mission. Assuming that the satellite is a Leo 500km satellite, it firstly carries the ADSB function to detect aircraft flight data. ADSB refers to the broadcast automatic correlation monitoring technology, which can provide aircraft flight position and flight status. Then, when the X area is reached, it is no longer necessary to detect the aircraft, but rather need to turn it into a signal processing monitoring module. Therefore, it needs to be reconfigured into a signal processing module. When running to the Y area, the signal processing mode changes, but it does not need to reconfigure all of it, only needs to change the processing method, so it only needs to reconfigure the signal processing part.

3.1. Mission management
The formula of orbital period is shown in (1), in which T_s is the orbital period, R_e is the earth's radius, h is the orbit altitude, G is the gravitational attraction constant, and M is the mass of the earth:

$$T_s = (\frac{(R_e+h)^3}{GM})^{1/2} \quad (1)$$

The satellite with a height of 500km has an orbital period of about 95 minutes. According to the experience in [6], the transit time only lasts for a few minutes. But it is enough to complete the mission and upload the configuration files in that time. Assuming that the reconfigurable bit file is stored in the file library, it is verified as the correct file. The following is a list of missions. The hardware area, execution time and space attributes and priorities of the mission are determined by the ground and sent to the satellite by the ground control station.

The area where the reconfigurable bit file was loaded is determined by the configuration file by addressing. A diagram is a specific address correspondence as in table 4. Through the communication channel described above, the reconfiguration mission can be mapped to the actual FPGA resource under the conditions of the presupposed address, and the control and interaction required for the mission, such as reading and writing data, will be completed. The size of the file in the table can also clearly distinguish entire reconfiguration and partial refactoring as in table 5.

Table 5. The address and the length of reconfigurable bit file.

Mission	Address	Length(hex)
ADSB	0x10000000	0098 9680
Add	0x02000000	3650
Product	0x01000000	3650

3.2. Configuration and implementation of reconfiguration
ZYNQ, the Xilinx Virtex series, is chosen in the paper. Its FPGA chip is a series of FPGA chips which can be implemented by Xilinx to realize dynamic partial reconfiguration. And the FPGA chip supports the application of space class, so this paper uses it to do experimental verification. ZYNQ belongs to ZYNQ-7000 series, and it adopts ARM+FPGA architecture. After system reset and successful startup, the first running software is BootROM (no disk startup ROM), loading and running FSBL (First Stage Boot Loader, FSBL) and user code to complete the system initialization configuration. The control of FPGA is done by PS, so after receiving the reconfiguration command of the FPGA, boot the bit file by the PS and implement entire or partial reconfiguration [13].

Two projects are designed in this paper. The first is the ADSB mission, and the second is the signal processing mission which can realize partial reconfiguration that has stated above. The satellite carries ADSB, which can monitor the aircraft in a wider range and faster. When running to the area where signal processing is needed, the ADSB function is reconfigured to perform signal processing missions. This partial reconfiguration can be applied to the update of signal processing algorithm and partial modification of signal modulation system.

For a specified part of the reconfiguration area, as long as the interface is designed, and the number of FPGA resources is enough in this area, we can design updates to any required processing. For example, the signal processing algorithm product module, which is designed in this paper, is a

reconfigurable module RM with two input signals and an output signal, which can be designed as a required algorithm in RM. This consideration is very practical for satellites. For example, for signal processing, the interface generally does not change much, but the processing algorithm will need upgrading after a period of time. The communication system will change in 3 or 5 years or others. As shown in figure 4, this article uses MATLAB/Simulink to carry out a design of addition and production. Through this tool, the automatic generation of Verilog code can be completed quickly and the reconfiguration of the two algorithms is completed. It is similar to the two orthogonal signals of I and Q paths in the BPSK signal. At this point, signal synthesis or orthogonality is completed, and then output. This is a design for partial reconfiguration. The design can complete part of the function replacement, can be used for software upgrading, interface to reduce power and so on.

3.3. Assessment

The assessment of real-time is divided into entire reconfiguration and partial reconfiguration. The first is entire reconfiguration. When the satellite function requires completely reconfiguration, it is usually the time to upgrade the satellite function or when the malfunction occurred and the satellite needs to be reconfigured. In that case, the satellite does not require very high real time. It should be noted that this paper does not consider the upload time of reconfigurable bit files. To calculate the time needed to upload files to satellites, we should consider the capacity and speed of the satellite link, and so on. This is not the factor of reconfiguration. The design scheme of this paper opens up a specific storage area for reconfiguration bit file, waiting for the storage of the bit files to be uploaded and checking the checksum. Only to make sure the reconfiguration bit file is correct, does the reconfiguration instruction will be sent and the satellite functions are entirely or partially reconfigured. Because the interruption of FPGA in the middle will affect the operation of the program, so the measurement of time is relatively rough. The reconfigurable bit file the paper use is 4261 KB and the time that from sending the command to completing the entire reconfiguration is about 5 seconds. After analysis, transferring the reconfigurable bit file from the on-board computers to the DDR takes up the most part of time.

Table 6 shows the average time the two partial reconfiguration (Product and Add) taken from receiving the reconfiguration instructions to completing the partial reconfiguration. The test method uses the PMU (Performance monitor unit, PMU) in the ARM kernel to implement [14]. For partial reconfiguration in real time, compared with entire reconfiguration, it has higher real-time requirements. This is determined by the different applications. In the case of partial reconfiguration, in addition to the fact that the requirements for the update of the function may not be very high in real time, other changes such as the replacement of the software radio wave form processing and the transformation of the I/O type are both a strong real-time transformation. As shown in table 6, we can see that partial reconfiguration time is very small, at 100 μs level, and it can be used in most occasions.

As for reliability, the paper discussed above is concerned in some ways. First of all, the on-board computer ensures the reliability of satellite to ground transmission. Then, the reconfigurable bit file library is stored in the storage area which is resistant to space irradiation, and the files all are checked and verified. When loading the file, firstly pass it to the DDR, then after the verification, do the transmission and then load it to FPGA. The "readback" technology can be used to compare the configuration data, all register contents, and storage numbers [15]. If FPGA resources are enough, three times redundant backup can be carried out, and vote circuit also can be used to make a judgement to improve the reliability.

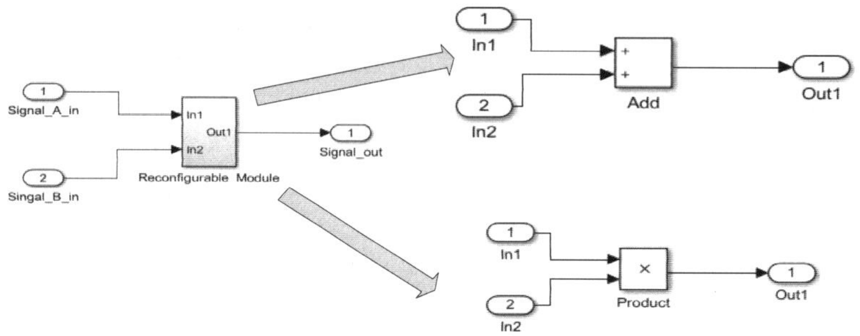

Figure 4. Partial reconfiguration to realize the upgrade of partial function.

Table 6. Real time test.

RM	Product	Add
Time/μs	1914	869

4. Conclusion

In this paper, a mission oriented spaceborne FPGA reconfiguration technology is introduced. Based on the region model, the problem of transformation from the reconfiguration mission to the execution of the hardware resources is solved. Not only the allocation and use of the global or local resources of the spaceborne FPGA can be realized through the technology of entire or partial reconfiguration, but also the use of FPGA can be effectively improved. Thus, the flexibility and autonomy of satellites can improve greatly. By assessing the real-time, reliability and security of the technology, it shows that this design scheme can be effectively used in mission oriented spaceborne FPGA reconfiguration.

References

[1] Y D Wen 2016 *Research and design of mobile communication system based on Software Radio.* Unpublished undergraduate dissertation, Hunan University, Changsha
[2] Eutelsat-Quantum : An overview of the European quantum satellite project http://www.360doc.com/content/17/0531/05/22081874_658600941.shtml
[3] S Shen, Z Y Zhu, J Wu 2014 *Reconfigurable hybrid mission scheduling algorithm. Computer application.* **32(2)** L387–390
[4] M Jie, J X Wang 2018 *A reconfigurable resource management model and its scheduling technology. Computer engineering and Applications.* 7 L 245–250.
[5] C P Xu, Y H Chai, W C Huang 2015 *Reconfigurable high performance mission scheduling system based on spatio-temporal context. Laboratory research and exploration.* **3** L116–121
[6] G liang, W B Liang, H J Liu 2010 *Design of reconfigurable communication system for low orbit satellite. Journal of aerospace.***31(3)** L185–191
[7] Brebner G, Diessel O 2001 *Chip-based reconfigurable mission management. Proceedings of the 11th International Workshop on Field-Programmable Logic and Applications.* **2147** L182–191
[8] Marescaux T, Andrei B, Dideriek V, Serge V, Rudy L 2002 *Interconnection networks enable fine-grain dynamic multi-missioning on FPGAs. Proceedings of the 12th International Conference on Field-Programmable Logic and Applications.* **2438** L795–805
[9] Busquere J P, Duearouge B, Grenier K 2005 *Above IC MEMS capacitors for integrated reconfigurable circuits. Microwave Conference Proceedings.* L4
[10] P Pang, W N Hao, W F Zhang, Y Xu, J B Zhu 2017 *An engineering implementation scheme for*

SRAM-FPGA in orbit reconfiguration. Spacecraft Engineering. **26(5)** L51–6

[11] Xilinx, Inc 2015 *Zynq-7000 All Programmable SoC Software Developers Guide*

[12] Xilinx, Inc 2015 *Reconfiguring Processor Peripheral Lab*

[13] J H Lu, Z Jiang, M Ma 2013 *Guidelines for hardware and software co design of embedded systems: Based on Xilinx ZYNQ* (Beijing: China Machine Press)

[14] *How to calculate the execution time of a certain program in SDK*: https://blog.csdn.net/wanruoqingkong/article/details/37759015.

[15] L Fang, S J Wu, L Min, Y S Gong 2014 *Translate. FPGA security design guide* (Beijing: China Machine Press)

IOP Conf. Series: Journal of Physics: Conf. Series **1195** (2019) 012013 doi:10.1088/1742-6596/1195/1/012013

Object-oriented Implementation of Chess Game in C++

Yiran Zhong

Chongqing Foreign Language School, Chongqing, China

Email: 779507224@qq.com

Abstract: In this paper, we implement the standard chess game using C++, a popular object-oriented programming language. Our program is developed and fully tested on Mac OS X system. It can be ran in terminal and allows two players to compete together. The object-oriented characteristics of C++ include namely, abstraction, inheritance, encapsulation and polymorphism are highly facilitates the development. At the end of the paper, we also points out some aspects that can be further improved.

1. Introduction

1.1 Background

Chess game belongs to casual games, and it requires two players to proceed on a chessboard of size 8 × 8 [1]. The rules of this game are transparent but have infinite possibilities to move pieces, thus each step is crucial to the player. The chess game is a highly adversarial game that requires players to consolidate step by step and develop their logical thinking ability. Computers, a machine that can perform numerical calculations, logical operations, memory storage, and program operations, are already playing an indispensable role in this digital age. It can replace human thinking and deal with problems that human beings can not solve. Computer provides chess game a brand new carrier, which not only makes it more convenient for people to play recreational games, train their intelligence and improve their chess skills, but also enables people to see the omnipotent compatibility of the computer field and the development vitality of other industries brought by computers. Even more it is helpful to promote research in the field of artificial intelligence [2]. The crucial role the computer plays in the chess game is storing a large number of game records and remnants, acting as assistants in the game, helping players check the game, studying opponents, preparing and analyzing the situation before the game. The methods of analyzing chess games and documenting rules presented by the computer may help solve many practical problems such as economic management, military command and so on in the future.

Our work uses C++ programming language to implement the chess game. With the help of object-oriented [3] properties of such language, designers are able to flexibly apply theoretical knowledge to solve practical problems, improve manipulative ability, and developers are encouraged to profoundly experience the basic thought—detailing the problem and decomposing things into various objects to solve—through the development of chess game. Compared with process-oriented C language, C++ is more suitable for large-scale software projects such as chess which includes multi-task, multi-factor and multi-requirement. By using encapsulation, we can encircle the process and data so that each functional module is independent of each other then we could successfully restrict the modification and maintenance of the program to the class, which ultimately makes things

easier. By applying inheritance, designers could share attributes from parent classes to improve code's reusability. By making use of polymorphism, it provides a way to abstract the complicated relationships. The scope of computer application is enlarging, and the scale of software is expanding. Designers urgently need to use more advanced design ideas like object-oriented to avoid the second "software crisis" [4].

1.2 Requirements analysis and settings
Our program enables two users to play chess, which means there is no AI involved in our work. The game needs to create different pieces, visualize the chessboard, accept various users input controls, and know when to terminate the game process.

- development environment: Mac OS X
- applicable environment: Mac OS X
- tools: sublime text, terminal

1.3 Roadmap
The rest of the paper is organized as follows. In Section 2, we discuss the object-oriented design for the program, especially the underlying abstraction. In Section 3, we explain the how to interact with our program in the terminal. In Section 4, we discuss several termination situations of the program. Finally, we conclude the paper with a short discussion about possible improvements in Section 5.

2. Game Settings and Abstraction
In chess game, the chessboard and the pieces are the essential elements. Without these two parts, all the functionalities cannot be achieved. Meanwhile, these two parts are closely related because the board must restrict the movement of pieces at any time, and the game cannot run without the pieces. To put all these into the nut shell, the design and initialization of chessboard and pieces should be given priority.

•Before talking about any design of the game, we need first to specify the settings of the game:

•The chessboard is two dimensional and has size 8 × 8.

•Players belong to two different camps, say white and black, and the white pieces move first.

•The chessboard is oriented, white is at the bottom of the board, and black is at the top of the chessboard.

•There are six types of pieces: king, queen, bishop, knight, rook, and pawn. Each Player starts with one king, one queen, two bishops, two knights, two rooks, and eight pawns.

•At the beginning of the game, the white pawns are in the second raw, and the black pawns are in the seventh row. Setting the rooks at the corners of the board, and the knights are close to the rooks, bishops near the knights, queen near the bishop. The rest square is king's position.

A standard chessboard should look like Figure 1. Next, we need to declare the movement rules of different pieces:

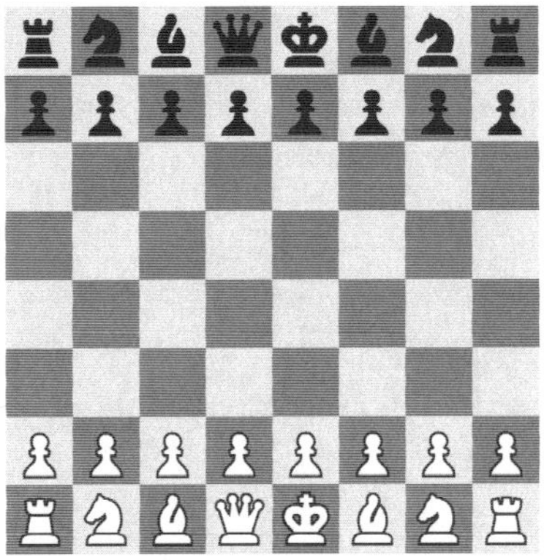

Fig 1: The standard setup of a chessboard [5].

•A king may move in any direction including diagonally, but may only move one square. A king may not move through other pieces. Also, the king cannot move into the square controlled by enemy's piece, otherwise, count as the foul. The king will be "checked" by other pieces, so we should code the special judgement for king.

•A queen may move any number of spaces in one direction, including diagonally. A queen may not move through other pieces .

•A bishop may move diagonally only, but may move any number of spaces. A bishop may not move through other pieces.

•A knight may move in an L-shape, of length either two-by-one or one-by-two. The knight is the only piece that is not stopped by other pieces in its way (i.e., it can move through other pieces to get to an open square).

•A rook may move any number of squares, but only in a straight line which is not diagonal on the board. A rook may not move through other pieces.

•A pawn can move only forward towards the opponent's side of the board, but with restrictions. On its first move of the game, a pawn may move forward either one or two squares; on subsequent moves, a pawn may only move forward one square. A pawn may not move through other pieces. Moreover, the pawn may not use this forward move to capture the opponent player's piece. Instead, the pawn moves one square diagonally-forward to capture a piece and have that piece removed from the board.

2.1 Board class and Chessgame class

Set up chessboard, create chess pieces, start games, move pieces, analyze chess game, and terminate games—these operations need to be abstracted as behavior of class objects. These functions must be implemented on the chessboard, so we are going to create a Board class to encapsulate all functions. The name of Board class indicates that it can be used in any games that are played on a board and the chess game was just a specific one of them. If we want to write some other board games (e.g., Chinese chess, checkers) in the future, we can directly inherit from the built Board class. This embodies the object-oriented languages' characteristics such as abstraction, inheritance, encapsulation and polymorphism.

The design of the Board class is as follows, where we only list some of its most important methods and attributes:

```
class Board {
public:
    /* Create a piece on the board.
        Returns true if the piece was successfully placed on the board. */
    bool initPiece(int id, Player owner, Position p);
    /* Return a pointer to the piece at the specified position,
        if the position is valid and occupied, nullptr otherwise. */
    Piece* getPiece(Position p) const;
    /* Perform a move from the start Position to the end Position.
        The method returns an integer with the status
        >= 0 is success, < 0 is failure.
        This method may handle the parts of moving pieces that
        generalize over different board games. */
    virtual int makeMove(Position start, Position end);
    /* The main gameplay loop. */
    virtual void run();
    /* Return true if the game is over. */
```

IOP Conf. Series: Journal of Physics: Conf. Series **1195** (2019) 012013 doi:10.1088/1742-6596/1195/1/012013

```
    virtual bool gameOver() const = 0 ;
protected:
    unsigned int m_width, m_height;
    /* Current game turn. */
    int m_turn;
    /* Vector containing all the Pieces of the board. */
    std::vector<Piece*> m_pieces;
    /* Functionality for creating a new piece (called by initPiece). */
    Piece* newPiece(int id, Player owner);
};
```

Successful implementation of above Board class helps us:

•create a new piece on the board by initPiece, which receives the ID of a piece, the holder, the position of the piece, according to certain rules, create the corresponding piece

•retrieve piece information at a specific location by getPiece, which receives the position, and returns the pointer of the corresponding piece if the position is legal

•move a piece on the board by makeMove, which is a very general method to move piece, i.e., put the piece in the ending position and then erasing the starting position of the piece

•start the main loop of the game by run, which shows the menu first, let the player choose to start a new game, and the system will initialize the chessboard and accept player's input to continue the game (details for interactions will be discussed in Section 3)

After we implement the Board class, we derive the chessGame class from it. It shares the methods and attributes built in the Board class, and then add more functionalities related to chess game. The design is as follows, where we only list some of its most important methods and attributes:

```
class ChessGame : public Board {
public:
    Position WhiteKingPosition;
    Position BlackKingPosition;
    /* Setup the chess board with its initial pieces. */
    virtual void setupBoard();
    /* Perform a move from the start Position to the end Position.
        The method returns an integer with the status
        >= 0 is success, < 0 is failure. */
    virtual int makeMove(Position start, Position end) override;
    void run() override;
    /* Return 0 if White is checked, 1 if Black is checked, -1 if neither. */
    int check();
    /* Return 1 if a stalemate is reached, 0 if not. */
    int Stalemate();
    /* Return 1 if a checkmate is reached, 0 if not. */
    bool checkMate(int status);
```

```
    /* Display the chessboard in terminal. */
    void printBoard();
...

protected:
    /* 1 if Black is checked, 0 if White is checked, -1 if neither. */
    int m_status;
};
```

It should be mentioned that we override the general makeMove method defined in Board class since we need not only require the movement of pieces in a general board, but also consider the specific rules that should be followed by each piece, which will be discussed in detail in Section 2.2. Moreover, during each move, we should care about the status of the game, such as whether there is a player that being checked, whether a checkmate or stalemate is reached and so on. All of these termination conditions should be incorporated in the overrode makeMove method of chessGame class and will be discussed in detail in Section 4.

2.2 Piece class and its derivations

Naturally, we also need to create a Piece class that abstracts the properties that an individual piece should have. The design is straightforward:

```
class Piece {
public:
    /* Return the owner of the piece. */
    Player owner() const;
    /* Return the id of the piece. */
    int id() const;
    /* Return an integer representing move validity, >= 0 valid, < 0 invalid.
       This method implements generic logic to check for a valid move. */
    virtual int validMove(Position start, Position end, const Board& board) const;
    /* Return 1 if the move is blocked by other pieces, 0 if not. */
    int handleBlock(Position start, Position end, const Board& board) const;
...

protected:
    /* Owner of a piece. */
    Player m_owner;
    /* Id of a piece. */
    int m_id;
    /* Constructs a piece with a specified owner. */
    Piece(Player owner, int id) : m_owner(owner) , m_id(id) {}
};
```

IOP Conf. Series: Journal of Physics: Conf. Series **1195** (2019) 012013 doi:10.1088/1742-6596/1195/1/012013

Note that Piece::validMove is used to deal with the generic logic to check for a valid move, which applies to all types of pieces, including determine whether the player moves the piece of the other side, whether the starting and ending positions of the piece are same, and whether the player capture the piece of its own side. If any one of them is true, an error will be returned.

In addition, all the pieces may be exposed to the problem that be blocked by other piece. Method handleBlock will help us analyze the movement from the horizontal, vertical, and diagonal directions. An error will be returned if the piece moves in these three directions and encounters other piece.

Since different types of pieces have different movement rules, the validMove is better to be overrode for different types of pieces. Therefore, we further derive the Piece class for all types and then override the validMove method in the derived classes. Following is the derived Pawn class from Piece class, and all the other types of pieces should be derived in a similar way:

```
class Pawn : public Piece {
protected:
    Pawn(Player owner, int id) : Piece(owner, id) {}
public:
    /* Implement piece-specific logic for checking valid moves. */
    int validMove(Position start, Position end, const Board& board) const override;
};
...
```

According to the rules specified before for each piece, the implementation for validMove in each derived class is not difficult:

•Pawn can only walk correctly in three situations: move forward one square, move forward two squares only at its starting position, move diagonally-forward one square to capture a piece of the other side. If none of these happens, then an error will be returned.

•For rook, after receiving the starting position and the ending position, we will check whether the horizontal and vertical coordinates of the starting position and the ending position are the same or not. If no one is the same, it indicates that the moving track of rook is not a straight line, and then an error will be returned.

•Knights are L-shaped moves, so we only need to consider two cases: 1: move one grid in horizontal way and then walk two grid vertically 2: move two grids horizontally and then walk one grid vertically. Return an error if player violates the rules above.

•Bishop can only move diagonally, so we only need to ensure that the number of squares that move horizontally after the movement equals to the number of squares that move vertically. If the player moves in other ways, an error will be returned.

•Queen can move in a straight line or in a diagonal line. This is just an combination of rook's rule and bishop's rule.

•The moving direction of king is the same as that of the queen, but king can only move one grid each time, so we need to ensure that the distance between the ending position and the starting position in each direction is one square. Otherwise, an error will be returned.

3. Interactions in Terminal

Throughout the game, column coordinates are specified with letters, and row coordinates are 1- based with numbers assigned starting from the bottom left . This is a standard chess notation, as is shown in Figure 1.

When starting the program, we will enter into a new game, the pieces will be set up automatically on the chessboard by appropriate constructor. And the turn counter will be initialized to 1. The turn

counter increases by 1 at every move, starting with turn 1 for white, moving on to turn 2 for black, then 3 for white, etc.

Table 1. (a) turn board mode on (b) white player moves one of pawns two squares forward (c) black player moves its- knight

row/column	a	b	c	d	e	f	g	h
1	Rook	Knight	Bishop	Queen	King	Bishop	Knight	Rook
2		Pawn	Pawn	Pawn	Pawn	Pawn	Pawn	Pawn
3								
4	Pawn							
5								
6						Knight		
7	Pawn	Pawn	Pawn	Pawn	Pawn	Pawn	Pawn	Pawn
8	Rook	Knight	Bishop	Queen	King	Bishop		Rook

(a)The arrangement of position

row/column	a	b	c	d	e	f	g	h
1	Rook	Knight	Bishop	Queen	King	Bishop	Knight	Rook
2	Pawn	Pawn	Pawn	Pawn	Pawn	Pawn	Pawn	Pawn
3								
4								
5								
6								
7	Pawn	Pawn	Pawn	Pawn	Pawn	Pawn	Pawn	Pawn
8	Rook	Knight	Bishop	Queen	King	Bishop	Knight	Rook

(b) The arrangement of position

row/column	a	b	c	d	e	f	g	h
1	Rook	Knight	Bishop	Queen	King	Bishop	Knight	Rook
2		Pawn	Pawn	Pawn	Pawn	Pawn	Pawn	Pawn
3								

IOP Conf. Series: Journal of Physics: Conf. Series **1195** (2019) 012013 doi:10.1088/1742-6596/1195/1/012013

row/column	a	b	c	d	e	f	g	h
4	Pawn							
5								
6								
7	Pawn	Pawn	Pawn	Pawn	Pawn	Pawn	Pawn	Pawn
8	Rook	Knight	Bishop	Queen	King	Bishop	Knight	Rook

(c) The arrangement of position

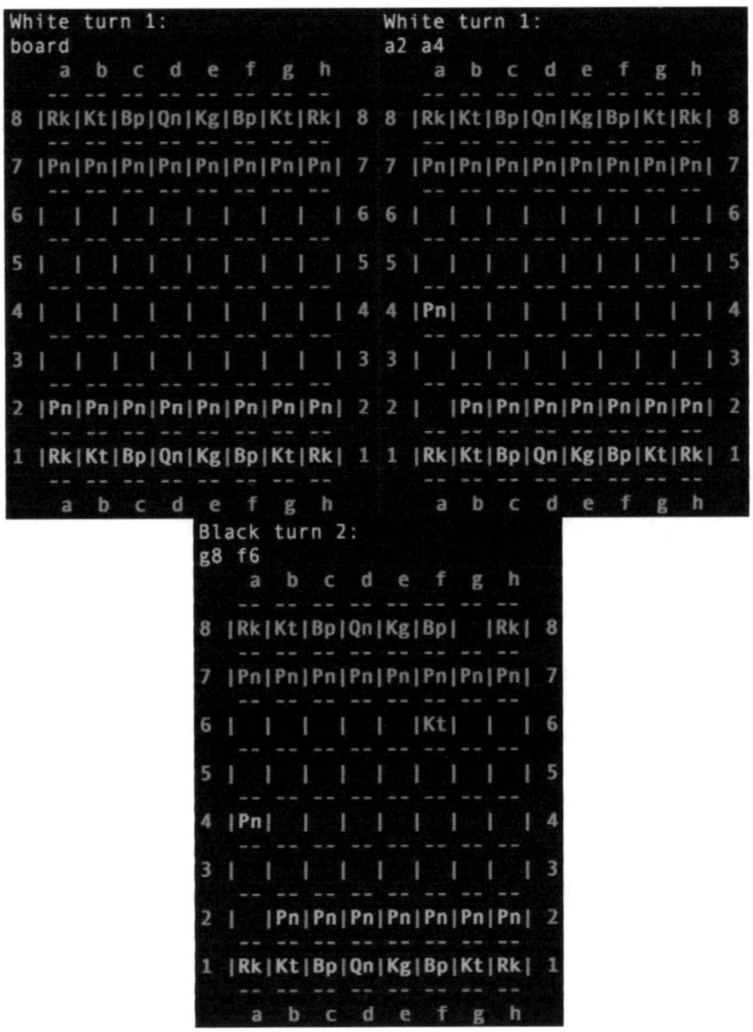

Fig 2: Illustration of some interactions in the terminal.

At the start of every turn, the program prints out a summary of the current state in the form of

<Player> turn <number>:

and waits for the user input. The program accepts a regular movement in a form of

<letter><number> <letter><number>

where the first letter/number combination stands for the starting position and the second letter/number combination stands for the ending position. For example, at the start of the game, it will print White turn 1:, and if the white player wants to move the leftmost pawn forward by two squares, we can type a2 a4. Certainly, for invalid positions, an error will be returned and the program continues to wait for a legit user input.

Besides the movement, our program also accepts some other non-move inputs:

•q: exit the game.

•forfeit: the current player forfeits the game so that the other player wins.

•board: toggles display of the board in the terminal. By default, the board is off. Entering board after being in no-board mode immediately displays the relevant board for the current turn. The board will now be shown after every turn, until board is entered again to turn the board off.

In order to be user-friendly, all of the above user inputs are case insensitive. And to display the board, we just produce a board similar to the standard chessboard using different color settings in terminal. An illustration of some interactions can be seen in Figure 2.

4. Termination Situations

This section focuses on several termination conditions of our program. The user-entered commands q and forfeit mentioned in previous section will exit the game, while there are also some other situations require the program to make judgment whether the game should be terminated.

If a king is ever threatened by any piece of the opposing player in such a way that a single move by the opponent could result in the king's immediate capture, this is considered a checked position. After each step, we have to decide whether this step will cause the opponent to be checked. The implementation is quite simple. After one step, use the enumeration method to find all of pieces of the current player, and check if there is any piece could capture the other player's king after moving one more step. If such a piece and corresponding pace exist, then it means that the opponent is in a checked position.

With the help of the ability to decide whether the opponent is being checked, we have two more termination situations of the game: checkmate and stalemate.

Checkmate means a player is checked but cannot get rid of this dangerous situation. Then the checking player is declared to be the winner of the game and the game should be terminated. We use a similar enumeration method to determine if a checkmate happens: traverse all the pieces of the checked side on the board to see whether there exists a pace that releases this checked position. If there exists such pace, then the checked player is not checkmated, and the game continues; if such a pace does not exist, then the game is over, and the checking player wins. It should be noticed that once a player is checked, the highest priority is to get rid of this position. And we must ensure that players would not voluntary to let king be in check by using enumeration.

Stalemate means a deadlock happens and each of the players is not able to take a step to checkmate the other. This can also be implemented by enumeration. When no one is in check, traverse the all the pieces of the player who has the right to move, and see whether the player has a legal movement. If there exists any legal movement, then this is not a stalemate and game continues; otherwise, the game is over and a stalemate message will be returned.

It should be mentioned that all of the above implementations involve using enumeration methods, where we will consider the movement of each piece one step in advance. However, this movement is not a real behavior of the piece. It only happens in our computation since we need to see its

consequence. Therefore, after each of such computations, we need to recover the board state to the one that before any further movements.

5. Conclusion

In this paper, we demonstrate our work of implementing the standard chess game with two players using C++. The properties of object-oriented language facilitate the development of this project. According to the standard rules, we make some high level abstractions of the game and then implement the details for each functionality. Meanwhile, various interaction ways in terminal are provided to make our program easier to used.

Of course, our work has many parts that can be improved. For example, add a scaler to record the number of pieces on both sides to make the whole situation clearer, or using drawing software to create the more refined chessboard. Also, the enumeration method used in determining termination conditions seems not to be optimal and we may use more advanced algorithms to enhance the efficiency.

This development experience is priceless, and it will help to lay a solid foundation for future learning.

References

[1] Chisholm, Hugh, ed. "Chess". (1911). Encyclopædia Britannica, 11th edition, Cambridge University Press (pp. 93–106).
[2] Russell, Stuart J. (2009). Norvig, Peter. Artificial Intelligence: A Modern Approach. Upper Saddle River, New Jersey: Prentice Hall.
[3] Bjarne Stroustrup. (1936).The C++ Programming Language. Addison-Wesley Professional, (4th ed).
[4] Edsger W. Dijkstra. (1972). The Humble Programmer. ACM Turing Lecture.
[5] https://www.chess.com/article/view/how-to-set-up-a-chessboard

IOP Conf. Series: Journal of Physics: Conf. Series **1195** (2019) 012014 doi:10.1088/1742-6596/1195/1/012014

A literature review on effects of time pressure on decision making in a cyber security context

Geoff Skinner[1,*] and Brandon Parrey[2]

[1]Advanced Cyber Security Engineering Research Centre, The University of Newcastle, Australia
[2]School of Electrical Engineering and Computing, The University of Newcastle, Australia

* Geoff.Skinner@newcastle.edu.au

Abstract. Shortages of time has become a natural characteristic of the professional environment. Individuals are often pressed to make fast decisions and complete tasks in a timely manner. This element of time pressure is particularly prevalent in the Information Technology (IT) sector due to fast-paced changes in demand, competition, and technology. Reviewing a wide range of journal articles, this paper aims to contribute to a broader discussion on the effects of time pressure on decisions. This paper explores the theoretical and practical considerations of decision-making, considering key decision-making models and the effect of technology on the decision-making process. Subsequently, this paper explores time pressure in a general sense, and then reviews the stress it causes individuals and its presence in technological environments, in particular a cyber-security context. After reviewing both decision-making and time pressure individually, the paper explores the relationship between the two, considering the influence time pressure on decision-making in technological settings. Concluding that time pressure can have an adverse effect on decision-making and hence possibly impacting cyber security services.

Index Terms—cyber security, technological environments, time pressure, decision making

1. Introduction

With increasingly complex environments at the hand of emerging technologies and competition, individuals and firms are often forced to make faster decisions (Salanova et al., 2002). Decisions are becoming more variegated, yet still require timely responses (Clark & Collins, 2002). Consequently, these external time pressures cause stress and fatigue to individuals (Bourne & Yaroush, 2003). This can affect their decision-making process and subsequently affects the output of their decisions. Despite the largely hegemonic acceptance that time pressure effects decision making, the impact in technological environments has been noticeably absent from the research agenda. Therefore, this paper aims to contribute to a broader discussion on the effects of time pressure on decisions by considering its presence and impact in technological environments.

Reviewing a wide range of journal articles from the fields of Engineering, Psychology, Management, Economics and Information and Communication Technology (ICT), this review will consider the theoretical and practical considerations of decision-making and time pressure. Subsequently, the relationship between the two concepts are considered, reviewing the influence of time pressure on decision making in technological environments. Specifically, in the context of cyber security, and the effect time pressured decisions might have on quality of service and support. Of most technological environments, cyber security is deemed the most critical real time service and subject to high pressure and need for quick decisions during crisis situations. For example, this might include the rapid spreading of a new virus in an Enterprise network, a Denial of Service (DoS) attack, or the detection of Trojan horses or real time data hacking activities.

2. Decision making in technological environments

Decision-making is typically defined as a mental process, which involves judging multiple options or alternatives (Bohanec, 2009). Making decisions usually requires evaluating at least two alternatives that differ on a number of attributes (LePine et al., 2005). Throughout the literature, there are many different decision-making processes, however there are two that have been widely accepted: rational decision-making and cognitive decision-making. Both theories follow three main steps: first, problem definition, secondly, identification, evaluation and selection of alternatives, and thirdly, implementation (Fredrickson, 1984; Mitchell & Beach, 1990; Clark & Collins, 2002; Baum & Wally, 2003). This review will consider the application of this process in both rational decision-making and cognitive decision-making.

2.1. Decision-making models

Whilst the existing literature concurs in the three steps of decision-making, the process of evaluating and selecting alternatives differs. Originating from the field of economics, the rational decision-making model is referred to as normative or prescriptive, where the decision problem is defined in terms of identifying the optimum decision, assuming an ideal decision maker who is fully informed (French, 1986; Bohanec, 2009). This method presupposes that complete information is available and that individuals are rational. Simon (1955) was the first to strongly criticize this view of perfect rationality, arguing that humans have a bounded rationality.

Simon's (1995) theory suggests that the limitations of the human cognitive system, as well as limitations in accessing relevant information, do not allow people to make perfectly rational decisions (Campitelli & Gobet, 2010). Rather, it is suggested that people 'satisfice', selecting a 'good enough' option, not necessarily the best option (Campitelli & Gobet, 2010,p.355). The notion that individuals are typically willing to settle for imperfect accuracy of their decisions in return for a reduction in effort is well supported (Bettman et al. 1990; Haubl & Trifts 2000). Experts in the field, Ariely (2008) and Haubl and Trifts (2000), suggest that individuals tend to use heuristics and take shortcuts to choose among alternatives in the decision-making process. In turn, the cognitive decision-making process has become more frequently acknowledged as the most realistic decision-making.

2.2. The influence of technology in decision-making processes

When dealing with the above mentioned decision-making criticisms, the question frequently posed throughout the literature is how can computers and information technology support people who are faced with difficult decisions, so that they can decide better, faster and more effectively (Bohanec, 2009). This is the main area of interest for computer scientists and information technologists, who try to provide effective methods and tools for supporting human decision makers (Bohanec, 2009). Many writers have suggested that the use of IT should improve strategic decisions and have developed a myriad of propositions about the effects of IT on strategic decision-making. Molloy and Schwenk (1995) examined the relationships between the use of computer-assisted information processing and strategic decision-making process and noticed that the use of information technology does improve

IOP Conf. Series: Journal of Physics: Conf. Series **1195** (2019) 012014 doi:10.1088/1742-6596/1195/1/012014

both the efficiency and, more importantly, the effectiveness of the decision-making process. Since then an extensive volume of literature has supported this notion.

Research on decision-making and technology has suggested that advances in computing and telecommunications technology have a positive impact on how people make group decisions. Baker (2002) explored this idea, comparing the performance of sixty-four virtual teams using different technologies. Bakers (2002) research presented that the introduction of collaborative video and audio technology significantly aided strategic decision-making within teams. Despite there being thorough research on the impact of technologies on decision-making, few writers have been able to draw on any systematic research into the effect of technology in decision-making in technical environments.

The benefits of technologies such as decision support systems (DSS) and communication technologies in decision-making are frequently discussed in the literature. It is proposed that these technologies support decision makers in a multitude of ways. Gonzalez and Kasper (1999) suggested that the primary goal for computerized decision support systems is to improve decision quality. This effectiveness has been reported in diverse settings. Demikan and Delen (2013) explored the benefits of DSS in a cloud environment. Their research shows DSS in cloud enables scale, scope and speed in organisations. Additionally, Bharati and Chaudhury (2004) investigated web-based decision support systems, finding that quality DSS were directly related to information relevance, accuracy, completeness, and timeliness.

However, this position on DSS is not unanimous in the literature. Some scholars argue that the use of DSS does not improve individual or group decision-making effectiveness. For instance, in a study of task complexity and DSS, Webby and O'Connor (1994) found that the DSS did not affect subjects' performance. Despite the findings about the role of technology in supporting decisions, research shows that technology can also complicate the decision-making process. According to Noorderhaven (1995), four cognitive factors influence the decision-making process1. Two of these factors are complexity and uncertainty. In the literature, the presence of these two factors is frequently discussed when considering technology.

Given the growth of information technology and the internet, an individual's ability to access large amounts of information has increased. Described as information overload, this is increasingly perceived as having an adverse effect on decision making by overwhelming the decision-maker and complicating the process (Bawden & Robinson, 2009). This was confirmed by a series of reports of large-scale surveys conducted on business managers by international news agency Reuter (Bawden & Robinson, 2009). Results showed that nearly half of the surveyed managers believed important decisions were delayed and adversely affected as a result of having too much information (Bawden & Robinson, 2009).

One criticism of these surveys is that it relies heavily on qualitative data. More comprehensive quantitative research is needed to make these studies more valid. Further, theory suggests that technology is the catalyst of uncertainty. It is argued that rapid technological change and turbulence in the marketplace creates uncertainty, and thus normal decision-making processes do not apply to technical environments (Baum & Wally, 2003). However, few studies test this hypothesis, and therefore further research is needed. The differing opinions in the literature suggest that there is still extensive research that should be done on the role of technology in decision-making.

3. Time pressure in technological environments

The shortage of time has now become a natural characteristic of the environment, with individuals often being pressed to complete increasing daily tasks in a short period (Klapproth, 2008). The reduction of time available to complete a task is referred to as time pressure (Klapproth, 2008). In the literature, the term time pressure tends to be used to refer to the psychological stress that results from having to perform duties in lesser time (Bourne & Yaroush, 2003). This is particularly prevalent in the IT sector, where individuals are frequently required to complete tasks and make decisions under significant time constraints. Broadly, the concept of time pressure is lacking in the literature. Rather literature explores the application of time pressure to real world situations - the effects and presence of

time pressure are explored. Therefore, the following part of this section considers in greater detail the effects of time pressure on individuals, and its growing presence in high technology environments.

According to LePine, Podsakoff, and LePine (2005), time pressure can have a beneficial effect on individuals, being associated with high motivation, productivity and improved performance. In one study, Cavanaugh, Boswell, Roeh ling, and Boudreau (2000) found that demands such as time pressure increase effort in individuals. Labeled as a 'challenge stressor' they argue individuals increase intensity in the hope to successfully cope with this demand and experience a sense of personal accomplishment or formal recognition (Cavanaugh et al., 2000). However, while being positively related to motivation and performance, more recent theory suggests time pressure increases strain and stress (Bakker & Demerouti, 2007; Syrek et al., 2013).

There is no single universally agreed definition of stress, however it is consistently reported that environmental conditions such as heavy workload and constant time pressure are major sources of stress and fatigue (Bourne & Yaroush, 2003). This is particularly the case in the IT sector, with IT employees ranking third as experiencing the most time pressure in a study conducted by the Federation of German Trade Unions (Syrek et al., 2013). Expanding on the 'challenge stressor' concept, Syrek, Apostel, and Anton (2013) explore the negative effects of time pressure in highly demanding IT jobs.

Time pressure is identified as a serious stressor, exhibiting a strong negative relationship to employee strain and wellbeing (Syrek et al., 2013). Additionally, scholars addressing employees in IT find that the level of exhaustion is particularly high due to the time constraints (Hetland et al., 2007). This stress and fatigue experienced due to time pressure causes individuals to deviate from normal cognitive and physical behaviour, thus affecting the decision-making process. The following sections of this paper will explore a more detailed account of time pressure in technological environments and consider how the stress it can cause individuals to experience subsequently affects their decision-making processes and outcomes.

Time pressure is increasingly prevalent in high-velocity environments, deriving from the need to make fast decisions and complete tasks in a timely manner. Particularly in the IT sector, because of globalization and increasing competition, individuals are under increasing pressure to meet deadlines and reduce response times (Salanova et al., 2002). High-velocity environments are described to involve fast-paced changes in demand, competition, and technology which may result in instability and uncertainty (Eisenhardt, 1989). When considering decision-making in these environments, Kedia, Nordtvedt and Perez (2002) claimed that traditional decision-making theories recommend individuals search for additional relevant information. However, many scholars have challenged this claim on the grounds that this delays the decision-making process and affects performance. Eisenhardt (1989) proposed a model of strategic decision-making speed for eight high-tech firms facing high-velocity environments. It was observed that the faster decision-making the better, with faster decision-makers having the best sales and profitability (Eisenhardt, 1989). Consequently, increasing time pressure on individuals. This theory, while important at the time, has become even more relevant to decision-making in the information age, with the importance of making timely decisions increasing with technology.

Decision-making in these high technology environments require quick decisions, as only timely decisions can lead to a competitive advantage (Audretsch, 2001; Clark & Collins, 2002). Consequently, decision makers in these environments often feel pressured to move quickly in order to be advantageous. Supporting Eisenhardt's evidence, Baum and Wally (2003) also concluded that faster decision-making has a positive impact on profitability and growth. However, by testing Eisenhardt's model on sixty-six high technology firms, they further concluded that environments with rapid changes in demand and discontinuous results, such as the IT industry, have more time pressures, in turn forcing a faster decision-making speed (Baum & Wally, 2003). This is due to decision speed being identified as a key enabler of technology firms to be able to exploit opportunities before they disappear (Baum & Wally, 2003). With increasing complex environments at the hand of emerging technologies, individuals and firms are often forced to make faster decisions. However, despite this

IOP Conf. Series: Journal of Physics: Conf. Series **1195** (2019) 012014 doi:10.1088/1742-6596/1195/1/012014

research widely exploring the idea that technology increases time pressure when making decisions, a more comprehensive study would have included the types of decisions made.

4. How time pressure affects decision-making

Emotion and time pressure are two important factors affecting decision-making (Hu et al., 2015). Individually, the effects of time pressure and stress are frequently discussed in decision making literature. However, despite the literature examining the influence of emotion and time pressure on decision-making, there is limited research considering the interaction of these two elements. Therefore, in the section that follows the effect of time pressure on decision-making, and the effect of time pressure induced stress on decisions will be considered.

Reviews of literature on time pressure have identified a number of ways in which the outcomes and processes of decision-making change when there is limited time available (Maule et al., 2002). The research agenda has been dominated by debates about how time pressure affects the quality of decision-making, with the general finding that individuals perform significantly worse under time pressure. Research into this area has been present in such fields as economics, management science, and psychology. However, there is a void in understanding the impact of time pressure specific to the field of IT.

Vance, Anderson, Kirwan and Eargle (2014) explored the effect of time pressure on critical security decisions. Participants responded to web browser security warnings when doing work under time pressure, thus simulating real-world working conditions. However, this study does not offer an adequate explanation of the effects of time pressure on the subject's decisions, as no attempt was made to quantify the association between the two. The study would have had higher validity if the author had placed a larger emphasis on the time pressure variable.

There are many computer-driven studies on the effect of time pressure on decision-making. Using computer simulations, many theories conclude that time pressure negatively effects the quality of decision-making. Studies consistently show that under time pressure, the informationprocessing strategy is altered. It is suggested under time pressure, people limit their information choices to that which can be attained quickly and easily (Savolainen & Kari, 2006). Tombros, Ruthven & Jose (2005) conducted a study that compared the information seeking behaviors of online searchers when assessing the relevance of web pages for information-seeking tasks.

Participants were given 15 minutes versus those given 30 minutes to complete search tasks (Tombros et al., 2005). They found evidence that the group with less time relied on more 'obvious' features such as query terms and link quality versus in-depth examination of content and structure (Tombros et al., 2005, p337). They also found that the 15-minute group reported the tasks as being more stressful, and they were less satisfied with their decisions (Tombros et al., 2005).

Even with the presence of decision supported technologies, time pressure is found to impair decision quality. An experiment was conducted by Smith and Hayne (1997), in which a group makes business decisions under time pressure, where half of the groups enjoyed the benefits of DSS and the other half had no access to a DSS. They found that the presence of time pressure was negatively related to quality of decision-making for both the supported and non-supported subgroups.

Decision strategies are influenced by emotion, which can play a bigger part in decisionmaking when people are under stress. One of the most obvious ways to put a performer under stress is to impose time limitations (Bourne & Yaroush, 2003). Conte (2015) experimentally explores the effects of time pressure on decision-making. Consistent with the results above, a significant proportion of participant's decisions were impaired by time limitations. Interestingly, this impairment is not due to the severity of the time pressure, but mainly due to the fact that being exposed to any level time pressure causes stress and panic in individuals (Conte, 2015). In turn, causing the subjects to use time inefficiently thus impairing decision performance. However, this research also exhibited a number of contradictory findings, with a number of subjects not affected by time limitations.

5. Cyber security time pressured decisions

As organizations are becoming more dependent on information technology, the emphasis on information security is getting more significant (Yayla, 2011). Consequently, the effects of time pressure and its related stress have been debated in the field of information and computer security. Whitman (2004, p49) considers the 'act of human error or failure' as one of the most severe threats to information security. Literature suggests that some of the underlying reasons behind user errors are due to time pressure and stress. West (2008) summarizes principles of decision-making in computer security.

Consistent with the previously discussed cognitive decision-making theory, he suggests people have limited time and mental resources when making decisions regarding online security. Consequently, West argues that the user is often motivated to get on with the primary task as quickly as possible and, therefore, less likely to make a pro-security decision (West, 2008). For instance, in cases where users are prompted to install software updates, scan a file for viruses before opening, and so forth, users are less likely to comply when in the middle of another task, especially if in a hurry (West, 2008). Further, Yayla (2011) considers the effect of time pressure related stress on individuals in information security. He proposes that user errors and negligence are arguably the two most common unintentional insider threats and that user errors are related to the 'unintentional threat' of time pressure (Yayla, 2011, p5). Yavla suggests that reducing work related stress and fatigue levels by adjusting time pressure and workload will reduce unintentional insider threats to information security. However, despite both Yayla (2011) and West (2008) investigating the effect of time and stress related variables on information and computer security decisions, there has been little attempt to quantify these theories. limitations.

6. Conclusion and future research

Due to continuous changes in competition and technology, the shortage of time is increasingly prevalent factor in decision-making (Klapproth, 2008). Despite the unanimous understanding that time pressure effects decisions, this literature review argues that the effect of time pressure on decision-making in technological environments has not been exhaustively engaged with in the literature. Upon the review of the literature, various gaps have been identified in existing research. When considering the role of technology in decision-making, there are differing opinions in the literature, suggesting that there is still further research needed. Additionally, the influence of time pressure and emotion on decision making is frequently discussed in the literature, however there is limited research considering the interaction of these two elements. This review suggests there is an adverse effect of time pressure and its related stress on decision making, negatively effecting the quality and process. This has been displayed in computer and web studies, and discussed in the field of cyber security. Both Yayla (2011) and West (2008) investigated the effect of time and stress related variables on online security decisions, however little attempt has been made to substantiate these theories. The authors and research center colleagues future research has a project centered on time pressure effects on Cyber Security decision making.

References

[1] S. Chen, B. Mulgrew, and P. M. Grant, "A clustering technique for digital communications channel equalization using radial basis function networks," *IEEE Trans. on Neural Networks*, vol. 4, pp. 570-578, July 1993.

[2] Ariely, D. (2008). Predictably Irrational. New York: HarperCollins Publishers.

[3] Audretsch, D. B. (2001). Research issues relating structure, competition, and performance of small technology-based firms. Small Business Economics, 16(1), 37-51.

[4] Baker, G. (2002). The effects of synchronous collaborative technologies on decision making: A study of virtual teams. Information Resources Management Journal, 15(4), 79.

[5] Bakker, A. B., & Demerouti, E. (2007). The job demands-resources model: State of the art. Journal of managerial psychology, 22(3), 309-328.

[6] Baum, J., & Wally, S. (2003). Strategic decision speed and firm performance. Strategic

Management Journal, 24(11), 1107-1129.

[7] Bawden, D., & Robinson, L. (2009). The dark side of information: overload, anxiety and other paradoxes and pathologies. Journal of information science, 35(2), 180-191.

[8] Beach, L. R., & Mitchell, T. R. (1990). Image theory-a behavioral-theory of decision-making in organizations. Research in organizational behavior, 12, 1-41.

[9] Bettman, J. R., Johnson, E. J., & Payne, J. W. (1990). A componential analysis of cognitive effort in choice. Organizational behavior and human decision processes, 45(1), 111-139.

[10] Bharati, P., & Chaudhury, A. (2004). An empirical investigation of decision-making satisfaction in web-based decision support systems. Decision support systems, 37(2), 187-197.

[11] Bohanec, M. (2009). Decision making: A computer-science and information-technology viewpoint. Interdisciplinary description of complex systems, 7(2), 22-37.

[12] Bourne Jr, L. E., & Yaroush, R. A. (2003). Stress and cognition: A cognitive psychological perspective.

[13] Campitelli, G., & Gobet, F. (2010). Herbert Simon's decision-making approach: Investigation of cognitive processes in experts. Review of General Psychology, 14(4), 354.

[14] Cavanaugh, M. A., Boswell, W. R., Roehling, M. V., & Boudreau, J. W. (2000). An empirical examination of self-reported work stress among US managers. Journal of applied psychology, 85(1), 65.

[15] Clark, K., & Collins, C. J. (2002). Strategic decision-making in high velocity environments: A theory revisited and a test.

[16] Conte, A., Scarsini, M., & Sürücü, O. (2015). Does time pressure impair performance? An experiment on queueing behavior.

[17] Demirkan, H., & Delen, D. (2013). Leveraging the capabilities of service-oriented decision support systems: Putting analytics and big data in cloud. Decision Support Systems, 55(1), 412-421.

[18] Eisenhardt, K. M. (1989). Making fast strategic decisions in high-velocity environments. Academy of Management journal, 32(3), 543-576.

[19] Fredrickson, J. W. (1984). The comprehensiveness of strategic decision processes: Extension, observations, future directions. Academy of Management journal, 27(3), 445-466.

[20] French, S. (1986). Decision theory: an introduction to the mathematics of rationality. New York:

[21] Halsted Press.

[22] Häubl, G., & Trifts, V. (2000). Consumer decision making in online shopping environments: The effects of interactive decision aids. Marketing science, 19(1), 4-21.

[23] Hetland, H., Sandal, G. M., & Johnsen, T. B. (2007). Burnout in the information technology sector: Does leadership matter?. European journal of work and organizational psychology, 16(1), 5875.

[24] Hu, Y., Wang, D., Pang, K., Xu, G., & Guo, J. (2015). The effect of emotion and time pressure on risk decision-making. Journal of Risk Research, 18(5), 637-650.

[25] Kedia, B. L., Nordtvedt, R., & Pérez, L. M. (2002). International business strategies, decision-making theories, and leadership styles: An integrated framework. Competitiveness Review: An International Business Journal, 12(1), 38-52.

[26] Klapproth, F. (2008). Time and decision making in humans. Cognitive, Affective, & Behavioral Neuroscience, 8(4), 509-524.

[27] LePine, J. A., Podsakoff, N. P., & LePine, M. A. (2005). A meta-analytic test of the challenge stressor– hindrance stressor framework: An explanation for inconsistent relationships among stressors and performance. Academy of Management Journal, 48(5), 764-775.

[28] Maule, A. J., Hockey, G. R. J., & Bdzola, L. (2000). Effects of time-pressure on decision-making under uncertainty: changes in affective state and information processing strategy. Acta psychologica, 104(3), 283-301.

[29] Molloy, S., & Schwenk, C. R. (1995). The effects of information technology on strategic decision making. Journal of Management Studies, 32(3), 283-311.

[30] Noorderhaven, N. (1995). Strategic decision making. Wokingham, UK: Addison-Wesley Publication.
[31] Salanova, M., Peiró, J. M., & Schaufeli, W. B. (2002). Self-efficacy specificity and burnout among information technology workers: An extension of the job demand-control model. European Journal of work and organizational psychology, 11(1), 1-25.
[32] Savolainen, R., & Kari, J. (2006). User-defined relevance criteria in web searching. Journal of Documentation, 62(6), 685-707.
[33] Simon, H. A. (1955). A behavioral model of rational choice. The quarterly journal of economics, 69(1), 99-118.
[34] Smith, C. A. P., & Hayne, S. C. (1997). Decision making under time pressure: an investigation of decision speed and decision quality of computer-supported groups. Management Communication Quarterly, 11(1), 97-126.
[35] Syrek, C. J., Apostel, E., & Antoni, C. H. (2013). Stress in highly demanding IT jobs: Transformational leadership moderates the impact of time pressure on exhaustion and work–life balance. Journal of Occupational Health Psychology, 18(3), 252.
[36] Tombros, A., Ruthven, I., & Jose, J. M. (2005). How users assess web pages for information seeking. Journal of the Association for Information Science and Technology, 56(4), 327-344.
[37] Vance, A., Anderson, B. B., Kirwan, C. B., & Eargle, D. (2014). Using measures of risk perception to predict information security behavior: Insights from electroencephalography (EEG). Journal of the Association for Information Systems, 15(10), 679.
[38] Whitman, M. E. (2004). In defense of the realm: understanding the threats to information security. International Journal of Information Management, 24(1), 43-57.
[39] Webby, R., & O'connor, M. (1994). The effectiveness of decision support systems: the implications of task complexity and DSS sophistication. Journal of Information Technology, 9(1), 19.
[40] West, R. (2008). The psychology of security. Communications of the ACM, 51(4), 34-40.
[41] Yayla, A. A. (2011, October). Controlling insider threats with information security policies. In ECIS (p. 242).
[42] J. U. Duncombe, "Infrared navigation—Part I: An assessment of feasibility," *IEEE Trans. Electron Devices*, vol. ED-11, pp. 34-39, Jan. 1959.
[43] C. Y. Lin, M. Wu, J. A. Bloom, I. J. Cox, and M. Miller, "Rotation, scale, and translation resilient public watermarking for images," *IEEE Trans. Image Process.*, vol. 10, no. 5, pp. 767-782, May 2001.

IOP Conf. Series: Journal of Physics: Conf. Series **1195** (2019) 012015 doi:10.1088/1742-6596/1195/1/012015

Robustness based low-energy multiple routing configurations for fast failure recovery

T Hatanaka[1,2] and T Tachibana[1,3]

[1]Graduate School of Engineering, University of Fukui, 3-9-1 Bunky, Fukui-shi, Fukui, 910-8507, Japan

[2]takayuki-h@network.fuis.u-fukui.ac.jp; [3]takuji-t@u-fukui.ac.jp

Abstract. Fast failure recovery and the energy consumption are important issues in communication networks. A low-energy based multiple routing configurations is proposed for realizing both the fast failure recovery and the low energy consumption. However, in this method, the network robustness and the performance of data transmission are degraded regardless of the low energy consumption. In this paper, for turning some link ports off while maintaining network robustness, we propose a robustness and low-energy based MRC by extending low-energy based MRC. Our proposed method utilizes an algorithm that determines some excluded links based on the network robustness. With the proposed method, the energy consumption can be reduced so as not to degrade the network robustness significantly. We evaluate the performance of our proposed method in some network topologies with simulation.

1. Introduction

Fast failure recovery is an important issue in communication networks because the communication networks such as the Internet are indispensable in our daily lives [1]. MRC (Multiple Routing Configurations) has been proposed as one of the fast failure recovery methods in [2]. In this method, multiple backup networks have been prepared for the failure recovery in advance.

The energy consumption is also one of the important issues for communication networks. ICT in 2030 may consume 51% of the global electricity supply [3]. In [4], MRC has been extended in order to both perform the fast failure recovery and reduce the energy consumption. With this method, the power consumption can be reduced by turning unused links off while performing the first failure recovery as is the case with MRC. However, the network robustness is degraded because all links that are not needed for the failure recovery are turned off.

In this paper, we propose a method called Robustness base Low-Energy MRC by extending [4]. Our proposed method utilizes an algorithm that determines some excluded links based on the network robustness. By using the proposed method, the energy consumption can be reduced while maintaining the network robustness. We evaluate the performances of our proposed method with simulation.

The rest of this paper is organized as follows. Section 2 introduces related work about low-energy based MRC and evaluation of network robustness. Section 3 explains our proposed method. Section 4 shows some numerical examples, and finally section 5 denotes conclusions.

2. Related work

2.1. Low-energy based multiple routing configurations

[4] has extended MRC so as to both perform the fast failure recovery and reduce the energy consumption. In this method, K backup configurations ($K \geq 2$) are designed previously as shown in figure 1. When a failure occurs at a single node or a single link, a network is switched to one of the K backup configurations. In each backup configuration, there are two kinds of nodes and four kinds of links. Nodes are classified into *non-isolated node* and *isolated node*, and links are classified into *normal link*, *isolated link*, *restricted link*, and *unnecessary link* according to those link weights (see figure 1). Restricted links are used to isolate nodes from traffic forwarding, isolated links are not used to transmit traffic, and unnecessary links are turned off.

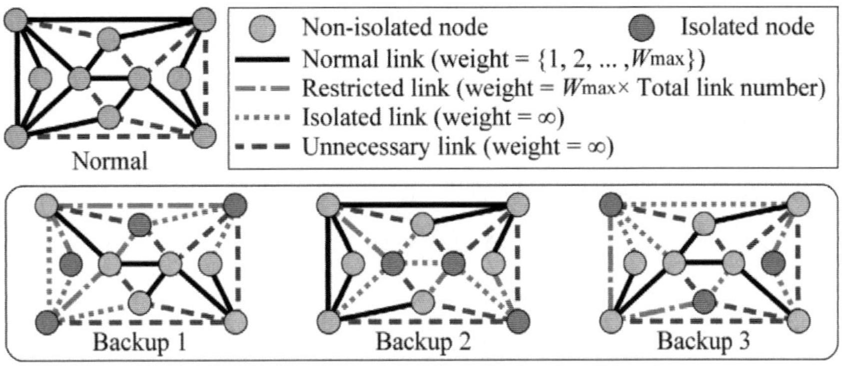

Figure 1. Normal configuration and backup configurations in a case of $K = 3$.

At first, in this method, K backup configurations are generated by using MRC. Then, a set L^* of links that are necessary for MRC is derived so that links in L^* satisfy the following two conditions:

(1) Each isolated node is connected to one or more restricted links.
(2) In each backup configuration, a connected graph is generated from all non-isolated nodes and all normal links.

When the failure occurs at a single node or a single link, it is possible to recover from the failure by using an appropriate backup configuration that does not use the failed node or the failed link.

2.2. Evaluation of network robustness with network criticality

As a metric for network robustness, network criticality has been proposed [5]. Now, let W be a weight matrix of a network and D be a diagonal matrix of weighted degrees for nodes that are in the network. From W and D, Laplacian matrix L is given by $L = D - W$.

In addition, let N be the number of nodes and n_i be the ith node n_i $(i = 0, \ldots, N)$ in the network. Let L^+ be a pseudo inverse matrix of the Laplacian matrix and L^+_{st} be the sth row and the tth column of pseudo inverse matrix L^+. Here, the network criticality τ_{ij} of link l_{ij} between n_i and n_j is given by $\tau_{ij} = L^+_{ii} + L^+_{jj} - 2L^+_{ij}$.

Now, let the amount of traffic from n_s to n_t $(s \neq t)$ be denoted as γ_{st}. In addition, the impact of the amount of traffic from n_i to n_j on the performance of the network is denoted as α'_{ij}. From the total amount of traffic $\gamma = \sum_i \sum_j \gamma_{ij}$, α'_{ij} is given by $\alpha'_{ij} = \gamma + (\gamma_{ij} + \gamma_{ji})/2 + \sum_k (\gamma_{ki} - \gamma_{ik})/N$.

Finally, network criticality τ' based on the amount of traffic is given by

$$\tau' = \sum_i \sum_j \alpha'_{ij} \tau_{ij}. \tag{1}$$

When the network criticality τ' is small (large), the robustness of the network is high (low).

3. Robustness based low-energy multiple routing configurations

In this section, we propose a Robustness based Low-Energy MRC so as not to degrade the network robustness by extending [4].

In this method, at first, unnecessary link set $\dot{L}$ is derived by using [4]. Then, the network robustness for the original communication network is evaluated by using a performance metric called network criticality τ' of subsection 2.2. Here, our proposed method utilizes new unnecessary link sets $\dot{L}_i$ ($i = 0, \ldots, |\dot{L}|$) that includes i unnecessary links. Moreover, the network robustness of the communication network where i links are excluded is denoted as τ_i' ($i = 0, \ldots, |\dot{L}|$).

In our proposed method, in order to exclude some links based on the network robustness, $\dot{L}_i$ are derived according to three steps. At the first step 1), $\dot{L}_0$ is set to $\emptyset$ and network criticality $\tau_0' = \tau'$ is calculated in a case where no link is excluded. Then, i is increased by one. Then, at the next step 2), for each link in $\dot{L} \setminus \dot{L}_{i-1}$, network robustness τ_i' is calculated for the communication network where all links in $\dot{L}_{i-1}$ and a link in $\dot{L} \setminus \dot{L}_{i-1}$ are excluded. Finally, at the last step 3), a link l is selected from $\dot{L} \setminus \dot{L}_{i-1}$ so that τ_i' becomes the minimum. Then, $\dot{L}_i = \dot{L}_{i-1} \cup l$ is derived.

Here, the threshold τ_{th} for the network robustness is given in advance and the network criticality is expected to be smaller than τ_{th}. Hence, from τ_{th} and τ_i', the proposed method determines the number ξ of links that are excluded according to the following equation.

$$\xi = \max_i \tau_i', \quad \text{subject to: } \tau_i' < \tau_{\text{th}}. \tag{2}$$

Then, all unnecessary links in $\dot{L}_\xi$ are turned off for reducing the energy consumption.

4. Numerical examples

In this section, we evaluate the performance of proposed method based on network robustness for the two network topologies in figure 2. In those topologies, the bandwidth for each link is 1,000 [Mbps] and the amount of communication traffic γ_{sd} between source node s and destination node d is 1.0 [Mbps].

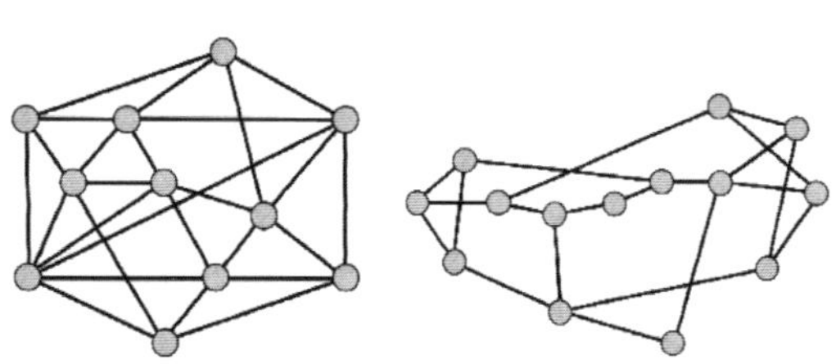

(a) COST239 (b) NSFnet.

Figure 2. Two network topologies.

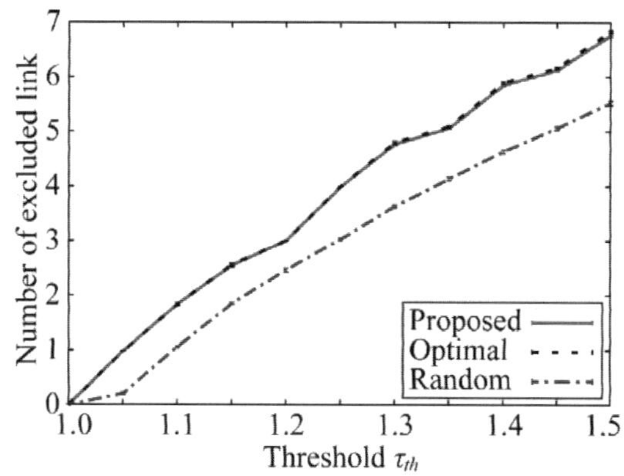

Figure 3. Number of excluded links for proposed method against threshold in COST239.

In the following, the proposed method is used to reduce the energy consumption by excluding unnecessary links while satisfying equation (2). For the performance comparison, we also consider a random method where unnecessary links are selected at random unless equation (2) is satisfied. We also consider an optimal method where unnecessary links are selected by using exhaustive search so that the energy consumption becomes the minimum while satisfying equation (2).

Figure 3 shows the average number of excluded links against the threshold τ_{th} for COST239. Note that the threshold τ_{th} is normalized by the network criticality τ'_0. From figure 3, we find that proposed method is effective for reducing the amount of energy consumption while maintaining the network robustness of network topology. This figure shows that the performance of proposed method is almost the same as that of the optimal method regardless of τ_{th}.

Moreover, figure 4 shows the number of excluded links against the number K of backup configurations for COST239. Here, the threshold τ_{th} of the network robustness is equal to 1.25 times the initial value τ'_0. From this figure, we find that the number of excluded links increases as the number of backup configuration increases. However, in COST239, the number of excluded links is almost constant when K is larger than three. This is because the network criticality dissatisfies the threshold τ_{th} if the number of excluded links increases by one. Moreover, we find that the number of excluded links for the random method is the minimum and the number of excluded links for proposed method is almost the same as that for the optimal method. From these results, proposed method can exclude a larger number of links according to the algorithm in Section 3 while maintaining the network robustness.

Figure 5 also shows the number of excluded links for NSFnet. In this figure, when K is smaller than five, the number of excluded links for proposed method is almost the same as those for other methods. When K is equal to or larger than five, the number of excluded links for the random method is the minimum. On the other hand, proposed method can exclude approximately the same number of links as the optimal method.

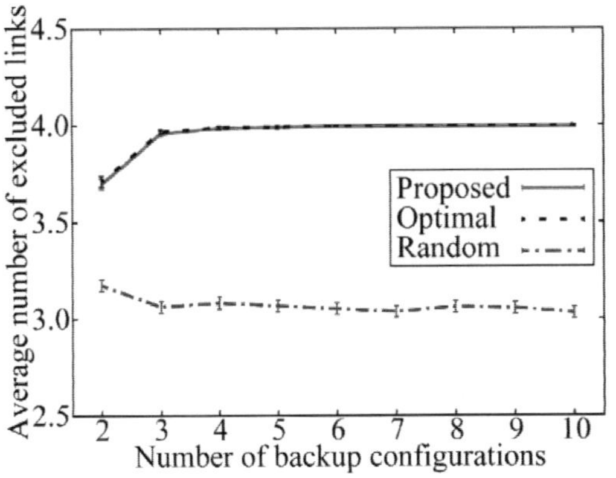

Figure 4. Number of excluded links against number of backup configurations in COST239.

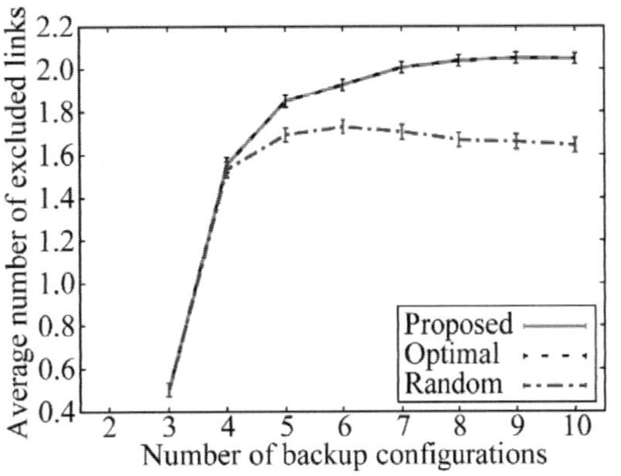

Figure 5. Number umber of excluded links against number of backup configurations in NSFnet.

5. Conclusion

In this paper, in order to maintain the network robustness in addition to realize fast failure recovery and low energy consumption, we proposed Robustness based low-energy MRC. We evaluated the performance of our proposed method in some network topologies with simulation. From numerical examples, we found that the amount of energy consumption can be reduced by using our proposed method while maintaining the network robustness. Therefore, our proposed method can realize fast failure recovery and low energy consumption. From numerical results, our proposed method are expected to be implemented and utilized in future communication networks by using MTR and SDN.

References

[1] Hoffman D L, Novak T P and Venkatesh A 2004 Has the internet become indispensable? *Communications of the ACM* **vol 47** no 7 pp 37–42

[2] Kvalbein A, Hansen A F, Čičić T, Gjessing S and Lysne O 2009 Multiple routing configurations for fast IP network recovery *IEEE/ACM Trans. Networking* **vol 17** no 2 pp 473–86

[3] Andrae A S G and Edler T 2015 On Global Electricity Usage of Communication Technology: Trends to 2030 *Challenges* **vol 6** no 1 pp 117–57

[4] Hatanaka T and Tachibana T 2015 Multiple routing configuration method for fast failure recovery and low energy consumption *2015 IEEE Int. Conf. on Consumer Electronics-Taiwan (ICCE-TW 2015)* June (Taipei, Taiwan) pp 478–79

[5] Fujihara T and Tachibana T 2014 Robustness evaluation and topology design based on amount of traffic for virtual network construction *IEICE Technical Report* vol 113 no 472 pp 13–8

Simplified automatic VAR/Power factor compensator using fuzzy logic based on internet of things

A N Luqman*, N S Lestari and I Setiawan

Department of Electrical Engineering, Diponegoro University, 50275 Jl. Prof. Soedarto, SH, Semarang, Indonesia

E-mail: alifnl@student.undip.ac.id

Abstract. In AC Power Systems, the compensation of reactive power is very important to support both of load and grid voltage. Generally, the objective of every reactive power compensators is to improve power factor that is the ratio of real power with apparent power to supplied the load. The main aim of this paper is twofold. Firstly, to design and realize a power factor corrector so the system's power factor is kept high, secondly to monitor energy consumed by the load using IoT technology. The uniqueness of this work is that to improve system power factor, we used capacitor bank which are composed from several capacitor with different values. The software is embedded in a low-cost microcontroller then will activate a unique combination of the capacitor by using successive approximation algorithm such that the power factor compensator more reliable; in this case, the damage of one or several capacitors in bank will not degrade the performance of the power factor compensator too much.

1. Introduction

Recently, many researches have been conducted about power factor correction unit, but still have deficiency because zero crossing detector circuit usage in the device [1] [2]. Local monitoring is still employed and causes inflexibility of power monitoring [3]. The main aims of this paper are: (1) designing simplified power factor correction unit, (2) reducing hardware existence by optimizing software ability, and (3) telemetry unit to monitor power usage in household electricity.

Various methods are applied to achieve the simplicity of the device. By using T/4 delay OSG (orthogonal signal generator), voltage and current magnitude can be obtained [4]. PQ transformation is also used to calculate active power and reactive power [5]. Fuzzy logic control is also used to estimate capacitance values. Our proposed device is also equipped by IoT technology, allows to monitor the system performance from anywhere. Auxiliary tools such as digital filter is also used such as: FIR (finite impulse response) and exponential filter for reducing noise that is produced by sensor. To get optimum execution time rate, Fuzzy, digital filter, and PQ transformation are implemented using fix point arithmetic.

2. Hardware setup

Figure 1 and figure 2 show proposed hardware's block diagram and hardware setup which had been built. Voltage sensor ZMPT101B and current sensor ACS712 are used to measure electrical network voltage and current, then read by using 12 bits embedded ADC in STM32f10c8 microcontroller. ENC28j60 ethernet module is connected to microcontroller through SPI (Serial Peripheral Interface) communication protocol; connecting the microcontroller module to the Internet network.

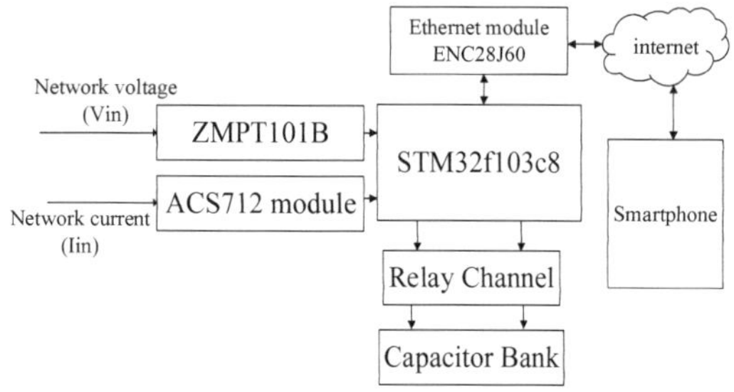

Figure 1. Proposed hardware diagram block.

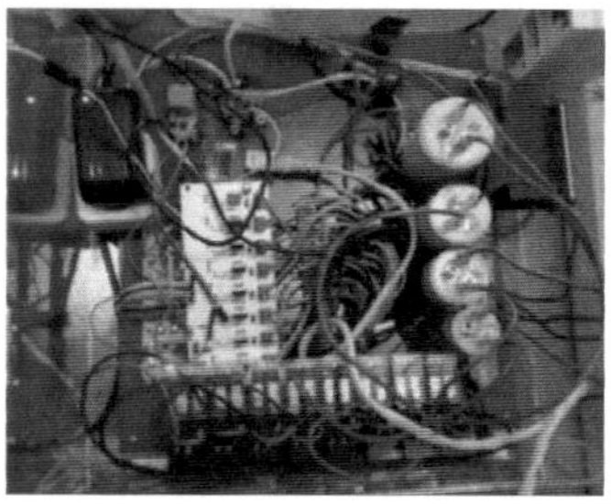

Figure 2. Realization of hardware setup.

Each capacitor is connected by relay driver to microcontroller. The capacitors in bank are represented by binary value (figure 3), thus the usage of successive approximation algorithm in the software.

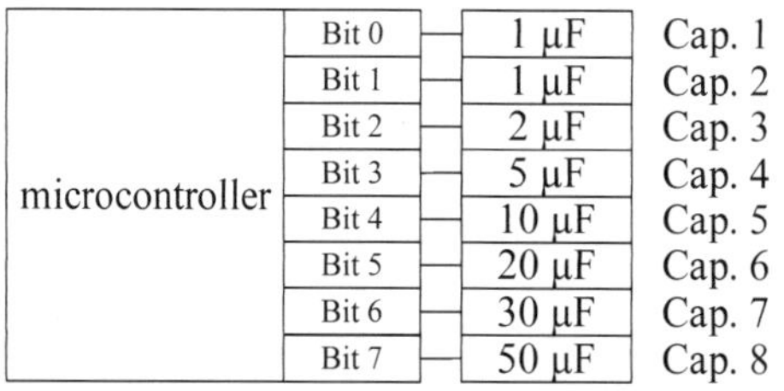

Figure 3. The capacitors' position in bank.

3. Software setup

Figure 4 shows software design for power factor correction unit. 12 bits ADC reads sensor analogue value and forwarded to 8th FIR digital filter. Active power (P) value and reactive value are obtained by PQ transformation calculation, whereas voltage and current magnitude are obtained by magnitude calculation method. Exponential filter takes place to smoothen computation results from magnitude and PQ transformation. Estimated capacitance values from fuzzy logic will be decomposed by successive approximation algorithm.

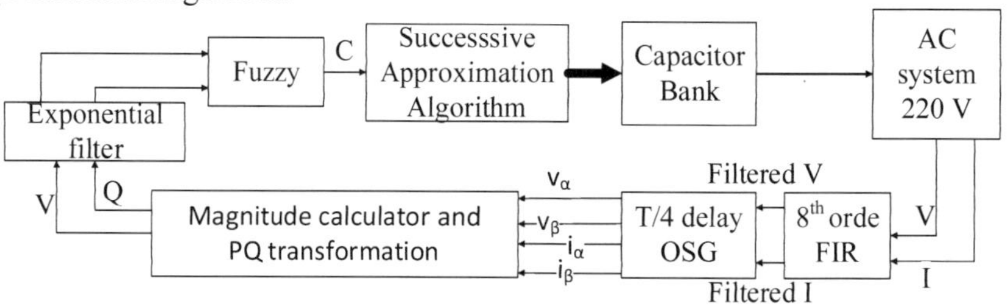

Figure 4. Software design in diagram block.

3.1. Digital filter

In this works, two types of digital filter are used; FIR and exponential filter provide smooth data to be used in other computations. FIR is designed with 7 delay units or 8th order. Equation (1) represents FIR 8th order formula, where x is filter output, u is filter input, and z is a delay unit.

$$X(z) = \frac{\left(1+z^{-1}+\ldots+z^{-7}\right)}{8}\, u(z) \quad (1)$$

Equation (2) represents exponential filter formula; it has a parameter that is α value in range from zero to one. The bigger α the less smooth filtering result, but filter's response is faster; the smaller α gives a smooth filtering result, but the filter's response is slower.

$$X(t)=\alpha\, u(t)+(1-\alpha)X(t-1) \tag{2}$$

Exponential filter was chosen because a very smooth result is desired. Using FIR maybe can use up to 60th order, which takes a lot of time for execution. Besides, by using exponential filter with 9.995 x 10^{-4} for α value within fix point computation method, execution time is more efficient and faster.

3.2. Magnitude measuring based on T/4 delay OSG

Equation (3) shows that magnitude of a sinusoidal signal which is delayed by ¼ period is equal to cosine value of the present angle. Amplitude or magnitude of a signal can be revealed by using trigonometry identity in equation (4).

$$\sin\left(\theta - \frac{\pi}{2}\right) = \cos(\theta) \tag{3}$$

$$\begin{cases} A(t)=\left((A_m \sin(\theta))^2+(A_m \cos(\theta))^2\right)^{\frac{1}{2}} \\ A(t)=A_m \end{cases} \tag{4}$$

Frequency sampling that used in this work in 10 kHz. For sampling both voltage and current signal, every period of 50 Hz sine wave there are 200 sampling data. In order to delay ¼ period of 50 Hz sine wave using 10 kHz frequency sampling, the signal must be delayed 50 samples with assumption the sine wave is steady at 50 Hz frequency.

3.3. PQ transformation

Reactive power and active power can be calculated by using equation (5).

$$P = \frac{1}{2}\,(v_\alpha \cdot i_\alpha + v_\beta \cdot i_\beta)$$

$$Q = \frac{1}{2}\,(v_\beta \cdot i_\alpha - v_\alpha \cdot i_\beta) \tag{5}$$

Where v_α and i_α are real voltage and current value, while v_β and i_β are T/4 period delayed value from voltage and current data or delayed by 50 samples in this case. Value of v_β and i_β are obtained from T/4 delay OSG.

3.4. Fuzzy design with ANFIS

Fuzzy is design by using ANFIS method with two input voltage, that is value and reactive power value; and an output is capacitor value. Voltage value has range 200 to 240 V, while reactive power value has range 0 to 900 VAR.

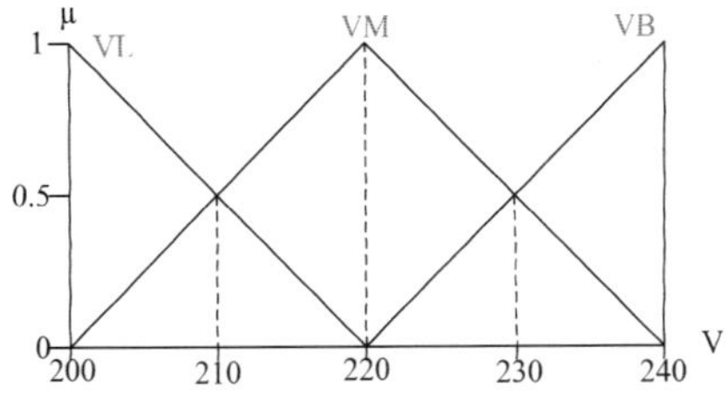

Figure 5. Membership functions for voltage input.

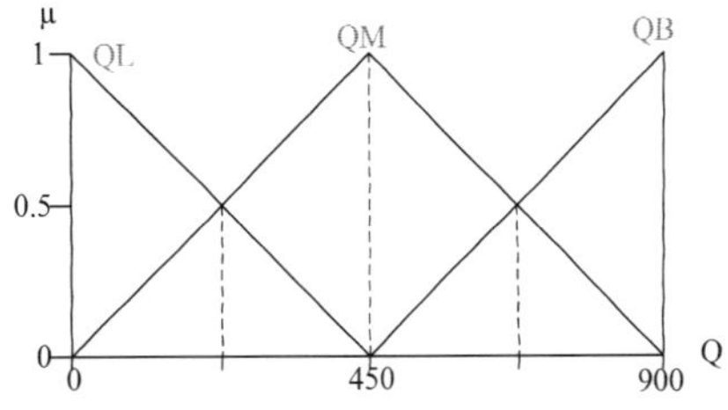

Figure 6. Membership functions for reactive power input.

As shown in figure 5 voltage magnitude is divided into three parts for fuzzification process. VL means voltage low. VM means voltage medium. VB means voltage big. Similarly, in figure 6 reactive power is also divided into three parts for fuzzification process. QL means reactive power low. QM means reactive power medium. QB means reactive power big.

IOP Conf. Series: Journal of Physics: Conf. Series **1195** (2019) 012016 doi:10.1088/1742-6596/1195/1/012016

Table 1. Fuzzy rule base after training.

Voltage input	Reactive power input	Output MF
VL	QL	w1
VL	QM	w2
VL	QB	w3
VM	QL	w4
VM	QM	w5
VM	QB	w6
VB	QL	w7
VB	QM	w8
VB	QM	w9

ANFIS training is done by MATLAB; after that, fuzzy rule set will be embedded to microcontroller as a usual fuzzy logic rule. The fuzzy rule base is shown in table 1. W1 until w9 mean the membership degrees weight after fuzzification process using "and" comparator. If the voltage and reactive power input are mapped, the capacitance values can also be mapped into 3D graphic as shown in figure 7. Increment of reactive power value will increase the capacitor; decreasing voltage value will increase the capacitor value and vice versa.

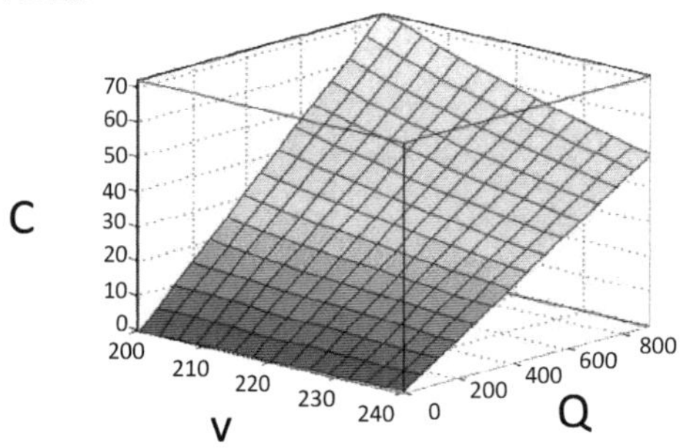

Figure 7. Fuzzy surface between input V and Q and output C.

3.5. Successive approximation algorithm

Generally used in ADC, this algorithm has a function to changes ADC input value to binary value. Based on its function, the authors chose this algorithm to energize and de-energize capacitors based on total capacitance must be activated. As an example, figure 8 will describe how this algorithm works. The profit of using this algorithm are faster and simpler computation than any other approximation methods [6].

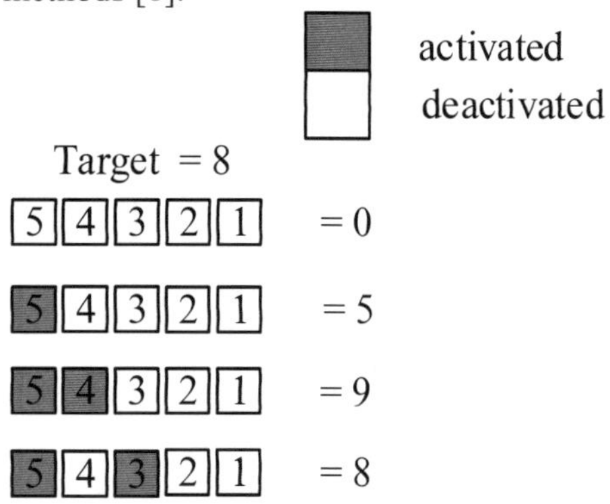

Figure 8. Successive approximation algorithm example.

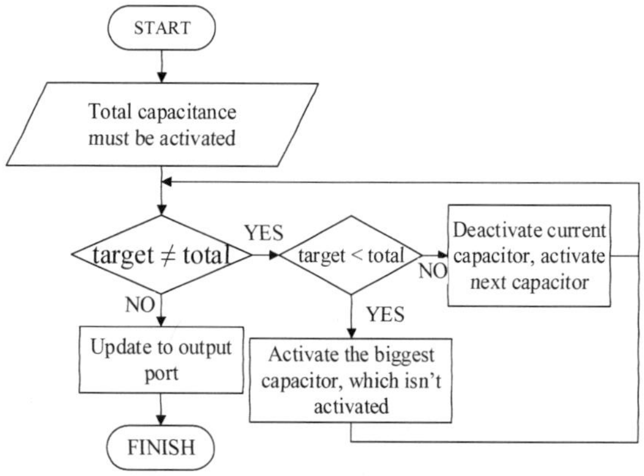

Figure 9. Successive approximation algorithm flowchart.

This algorithm will activate the biggest value first. In case, the present value is less than target value, then this algorithm will activate the next value. But if the present value is bigger than target value, this algorithm will deactivate recent activated value and activate the next value. This algorithm will iterate until the total activated value is equal to target value. For further explanation, figure 9 shows successive approximation algorithm flowchart, which has been used in this works.

3.6. Internet of things (IoT) setup

Message Queue Telemetry Transport (MQTT) is one of protocols for IoT for telemetering purpose. In this work, STM32 microcontroller will send data that contain power factor value, power usage value, and active capacitor as shown in figure 10. MQTT server that is used in this work is Thingspeak.com, for trial purpose ThingSpeak can be accessed and login directly to ThingSpeak website www.thingspeak.com. Data will be sent every 15 seconds.

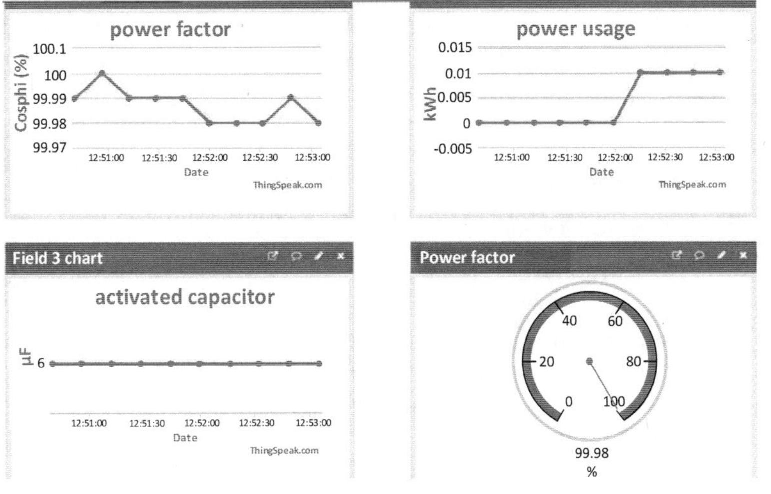

Figure 10. Thingspeak user interface.

4. Experimental results and discussions

This section will show all experimental results and will be analysed. Various load type has been chosen to observe the performance of this device, such as: various fan type, lighting bulbs, and induction motor. The load will be activated individually and together to check the precision of this device. Load type experiment which shown in table 2 and figure 11 is one of experimental results with 2 motors load with 0.25 kW power.

Table 2. Various load type test.

Load type	Power factor		Active capacitor	current (A)	
	Before	after		before	after
Fan I	1	1	-	0.17	0.17
Fan II	0.842	1	Cap. 3	0.18	0.13
Fan III	1	1	-	0.21	0.21
Fan IV	0.732	1	Cap. 1 & 3	0.26	0.16
Fan I, II, III, IV, and V	0.874	1	Cap. 1 & 4	0.79	0.65
2 light bulbs 200W	1	1	-	1.64	1.64
All fans + 2 light bulbs	0.994	1	Cap. 1 & 4	2.43	2.4
0.1 kW Motor I	0.851	1	Cap. 3, 4, & 5	0.27	0.2
125 W Motor II	0.475	1	Cap. 4, 5, & 6	1.28	0.21
Motor I and Motor II	0.514	0.936	Cap. 6 & 7	1.59	0.56

Figure 11. Result with 2 motor 0.25 kW load.

Inductive load and resistive load are installed on the electrical network together and discretely. Based on table 2, the device can compensate reactive power and maintain the power factor at around unity value. This device measured reactive power and remove it by activating capacitor based on the needed capacitance value. This device also can rebuild the needed capacitances value from capacitors bank precisely. Active capacitor in table 2 is capacitor number in figure 3.

Beside compensating reactive power from the network, this device also can perform power usage monitoring by using Internet of Things technology, as shown in figure 10. The poverty of using Internet of things technology is the power cannot be monitor in hard real time. There is a delay time for each data to be sent to Thingspeak's server.

5. Conclusion

This project provides a simpler hardware design to overcome power losses and power usage efficiency in household. AC voltage and current is measured by T/4 delay OSG without using any other auxiliary circuit. PQ transformation calculates active power and reactive power shorter. Needed capacitance values can be estimated using fuzzy logic controller. Designing fuzzy logic with ANFIS method eases the authors at designing fuzzy logic. The software is embedded to low-cost microcontroller. Telemetering function also works properly to send various data to MQTT server in interval 15 seconds. Overall the main function of this project that is to maintain unity power factor value in household electricity can be achieved using these methods.

References

[1] Khan MB and Wahab A. Automatic Power Factor Correction Unit. *IEEE*. 2015;(1).

[2] Biswas R Sen and Mal S. Automatic power factor improvement using microcontroller. *Int Conf Work Comput Commun IEMCON 2015*. 2015;

[3] Kabir Y, Mohsin YM, Khan MM. Automated Power Factor Correction and Energy Monitoring System. *IEEE, Trans Power Syst*. 2017;(February):5.

[4] Setiawan I, Andromeda T, Facta M, Handoko S. Implementation and Performance Analysis of a Single Phase Synchronization Technique based on T / 4 Delay PLL. *International Journal of Renewable Energy Research. 2018*;**8**(1).

[5] Yang Y, Blaabjerg F, Zou Z. Benchmarking of grid fault modes in single-phase grid-connected photovoltaic systems. *IEEE Trans Ind Appl*. 2013;**49**(5):2167–2176.

[6] Rodrigues GS, Kastensmidt FL, Pouget V, Bosio A. Exploring the Inherent Fault Tolerance of Successive Approximation Algorithms under Laser Fault Injection. *IEEE*, 2018;1–6.

Acknowledgments

The authors would like to acknowledge the support from the Energy Conservation Laboratory, Department of Electrical Engineering, Diponegoro University, Semarang, Indonesia. The authors also appreciate Joshua Harbangan for his helping hand. The research program RPI with project number 385-88/UN7.P4.3/PP/2018, which is financed by Univeristas Diponegoro, Semarang. This research is also supported by KEMENRISTEKDIKTI, Indonesian government under PKM-KC 2018 project.

Small displacement detecting method based on multifrequency continuous wave radar system

Aloysius Adya Pramudita*, Dharu Arseno and Erfansyah Ali

Telecommunication Engineering Department, Telkom University, Bandung, Indonesia

Email: pramuditaadya@telkomuniversity.ac.id

Abstract. A small displacement becomes important indicator in identifying a problem that may rises in several systems. Non-Contacting sensor is needed for small displacement detection in several field such as structure health monitoring, landslide monitoring and human vital sign detection. Radar system is potentially implemented as non-contacting sensor for previous mention problems. Detection of small displacement on a target using a radar system requires high accuracy and resolution which gives the consequence of a wide bandwidth requirement. The method of multi-frequency continuous wave radar is then investigated and proposed as a method for detection of small displacement with low bandwidth requirements. Cross correlation and IQ demodulation techniques are applied in post processing. Theoretical analysis and simulation were conducted to study the concept of the proposed method. The results show that the proposed method can determine the location and magnitude of small displacement.

1. Introduction

Small displacement in millimeter or centimeter scale is observed to identify a problem that may arise in several fields such as structure health monitoring, landslide monitoring and detection of human vital signs. For several cases, a non-contacting sensor is more precise to be implement than contacting sensor in detecting small displacement. Contacting sensor have some practical difficulties when to be used in detecting a vibration on boiler tank or large mechanical structure. For this case, the development a non-contacting sensor based on reflection coefficient measurement was studied [1]. However, the method can't be used to measure at a large distance. The non-contacting sensor must be placed near the target. The radar system potentially to overcome the measurement distance problems in previous method. Radar system has also been studied in developing non-contacting sensor in many problems such as structure health [2], landslide [3] and human vital signs monitoring (respiration, heart beat) [4, 5]. The small displacement is used as indicator of the problem that may rise on the previous mention cases. For small displacement in a millimeter or centimeter scale, high resolution radar is required. Wider bandwidth of radar signal is required for high resolution detection in the conventional radar point of view. In the other side, the complexity will rise in realization the radar system with wider bandwidth and the interference problem possibility should be carefully considered in the radar operation.

Processing of phase data of radar signal has been studied as a method for high accuracy detection. However, the phase repetitions for a target distance which larger than wavelength becomes limitation in detection capability that need to be overcome. There are several radar technologies that can be used

in detecting object including impulse, continuous wave (CW) and frequency modulated continuous wave (FMCW) [6]. FMCW radar has widely studied for many applications. Several methods for improving the radar resolution by enhancing the spectral computation [7-9]. However, it still difficult to achieve the detection accuracy for small displacement cases. CW radar is the simplest radar system which requires lower bandwidth in comparison with impulse and FMCW. However, the CW radar is only capable to determine the speed of movement but can't determine the distance or displacement. Impulse and FMCW have more complex system that require a larger bandwidth. Considering the advantage of CW radar that requires a smaller bandwidth, modify CW radar system to develop its capability in detecting a displacement will give a significant contribution in the development of radar system for small displacement detection.

This paper proposes a modification of CW radar to establish its capability in detecting a small displacement. The modification is done by generating multi frequency sinusoidal signal in transmitter side and developing appropriate post processing in software side. The results showed that the multi frequency CW radar system that proposed could detect small displacement in millimeter scale.

This paper is organized as follows. The first section discussed about the background that motivates the proposed method and problems that addressed in this research. The second section described theoretical perspective of the multi frequency CW radar that proposed for small displacement detection. The third section contains an analysis of the simulation results undertaken to assess the ability of the proposed system and the last is the conclusion.

2. Proposes multi-frequency CW system

The CW radar basically is consisted of sinusoidal signal generator, power splitter, power amplifier and transmitting antenna at transmitter side and low pass filter, mixer, low noise amplifier and receiving antenna at receiver side. Block diagram of CW radar system is shown on figure 1. CW radar system transmits electromagnetic waves with a single frequency of f_1 and target distance to the radar will affect the phase shift in the reflected wave that received by radar. If the distance between target and radar is d_0, then the transmitted waves that reflected by the target will arrive at the receiver with phase shift of $4\pi d_0/\lambda_1$. λ_1 is the wavelength of transmitted waveform. The received waveform can be written as (1). The reflected wave is then mixed with the transmitted signal in the mixer. Furthermore, the component of phase difference from mixer output is selected using lowpass filter (LPF) and the LPF output is the dc signal that related to phase shift (2). When a small displacement (d_t) ccurred on the target, the target distance to the radar and corresponding phase shift change respectively to $d_0 + d_t$ and $4\pi(d_0 + d_t)/\lambda_1$. Furthermore, the LPF output can be written as (3).

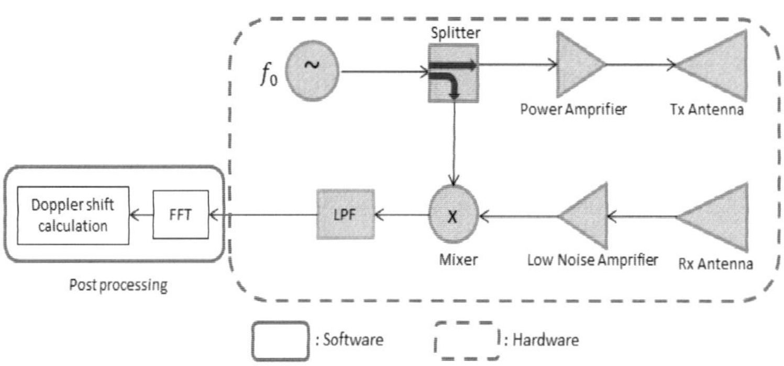

Figure 1. Block Diagram of CW radar system.

Information on target distance and is difficult to estimate using phase data in (2) and (3) because the phase repeats periodically corresponding to the wavelength of the CW signal. The distance information that allows to be estimated with phase data is a limited shift in the wavelength range.

However, the output of LPF such as (2) and (3) is a dc value consisting of amplitude and phase data which difficult to be extracted.

$$S_{RX}(t) = A_R \cos(2\pi f_1 t + \emptyset_1 + \frac{4\pi d}{\lambda_1}). \tag{1}$$

$$S_{LPF}(t) = A_L \cos\left(\frac{4\pi d_0}{\lambda_1}\right). \tag{2}$$

$$S_{LPF}(t) = A_L \cos\left(\frac{4\pi(d_0+d_t)}{\lambda_1}\right). \tag{3}$$

This paper proposes a several modifications of CW system, both in hardware and post processing parts. In order to overcome the phase repetition problem in CW output, the multi frequency sinusoidal signal generator is employed at transmitter side. The non-periodic signal then can be formed by combining the multi frequency sinusoidal signal. LPF2 is added at receiver side to extracted reference signal for demodulation and cross-correlation calculations. The block diagram of the multi frequency CW radar that proposed in this paper is depicted in figure 2. IQ demodulation technique usually was employed in estimating the phase of signal. Therefore, the IQ demodulation technique is used to estimate the phase data. However, in several previous researches, the IQ demodulation is located at radio frequency (RF) part [10, 11]. To avoid the significant modification in hardware side, we proposed performing the IQ demodulation in postprocessing side.

The proposed method uses a multi-frequency sinusoidal signal generator on the transmitter, but only one of the frequencies is used as a reference signal for the mixer process at the receiver. The signal that chosen as the reference is the lowest frequency sinusoidal signal. Multi-frequency CW radar signal which is then transmitted from the transmitter is then written as (4) where N, A_n and $\theta_n$are respectively as the number of frequencies used, the amplitude of each sinusoidal signal and he initial phase of each sinusoidal generator. The transmitted signal that reflected by the target and received by the receiver can then be written as (5) with λ_n is the wavelength of corresponding to each sinusoidal signal. Afterward, the signal is then mixed with a reference signal and the mixer output can be written as (6). The frequency difference components are then selected using LPF1 and the output of LPF1 can be expressed as (7). The LPF1 output is a sum of sinusoidal signals whose frequency is the frequency difference between each sinusoidal signals and reference signal. In each element of the sinusoidal signals there is a phase shift that corresponding to the distance between the target and the radar (d).

$$S_{TX}(t) = \sum_{n=1}^{N} A_n \cos(2\pi f_n t + \emptyset_n). \tag{4}$$

$$S_{RX}(t) = \sum_{n=1}^{N} B_n \cos(2\pi f_n t + \emptyset_n + \frac{4\pi d}{\lambda_n}). \tag{5}$$

$$S_M(t) = \begin{matrix} \sum_{n=1}^{N} C_n \cos(2\pi(f_n + f_1)t + \emptyset_n + \frac{4\pi d}{\lambda_n}) + \\ \sum_{n=1}^{N} D_n \cos(2\pi(f_n - f_1)t + \emptyset_n + \frac{4\pi d}{\lambda_n}) \end{matrix}. \tag{6}$$

$$S_{LPF1}(t) = \sum_{n=1}^{N} E_n \cos(2\pi(f_n - f_1)t + \emptyset_n + \frac{4\pi d}{\lambda_n}). \tag{7}$$

$$S_{Loopback}(t) = \sum_{n=1}^{N} F_n \cos(2\pi(f_n - f_1)t + \emptyset_n). \tag{8}$$

$$r_{LR}[k] = E(S_{LPF1}[k].S_{Loopback}[l-k]) \tag{9}$$

$$S_{LPF2}(t) = G_n \cos(2\pi(f_2 - f_1)t + \emptyset_n + \frac{4\pi d}{\lambda_n}). \tag{10}$$

Switch that connecting between the transmitter and receiver sections are used to determine the loop back signal which is then stored and used as reference in determining the initial distance (d_0)between the target and radar. The initial distance is determined by performing calculation of the cross

correlation between the loop back signal and the target signal as written in (9). Afterward, the target distance can be estimated from the peak position of normalized cross-correlation result.

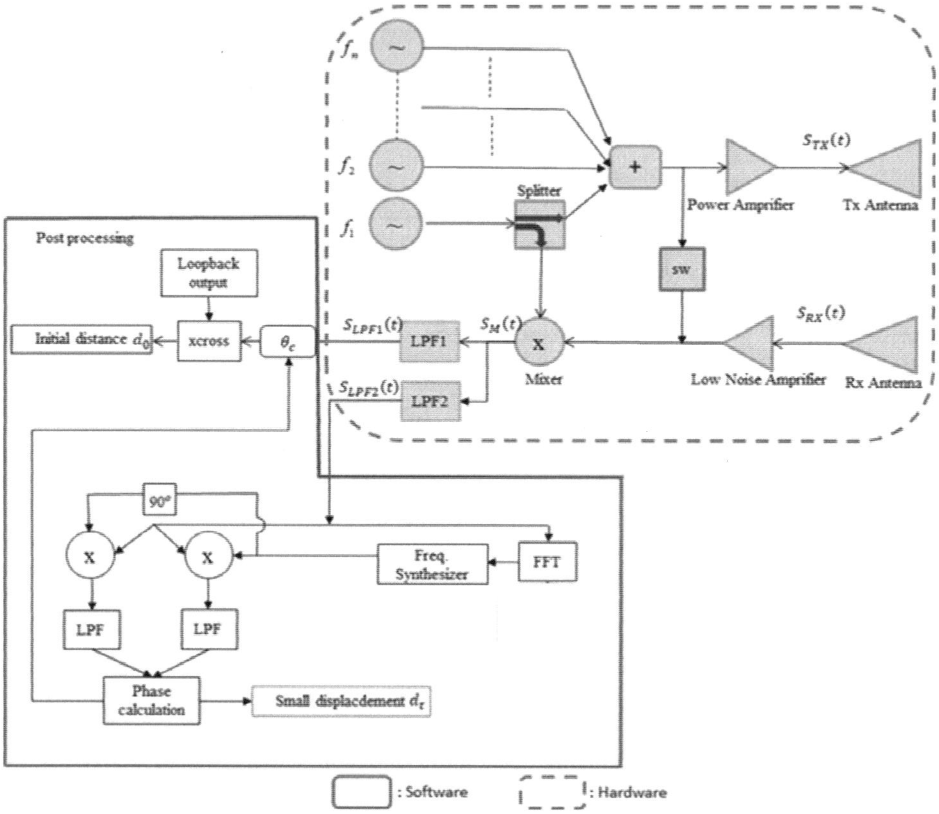

Figure 2. Block Diagram of proposed Multi-frequency CW radar for small displacement detection.

The displacement that may occurred on the target is determined from the phase data of LPF2 output signal. The phase data of LPF2 output can be determined by applying IQ demodulation processing that also discussed in [10, 11]. A sinusoidal signal with the same frequency as low pass filter output is needed in IQ demodulation method. After detecting the frequency of low pass filter output using FFT in numerical computation domain, then the reference sinusoidal signal that needed in IQ demodulation is easily synthesized. Further, the low pass filter output then mixed with sinusoidal signal that has synthesized before.

The theoretical analysis that discussed in this section indicated that the proposed method has capability to estimate the location (d_0) and small displacement (d_t) that happen on the target. The proposed method, basically is a modification and improvement in CW radar system to make the CW radar system has an ability to detect small displacement while maintaining low bandwidth.

3. Results and analysis

Computer simulation was performed to verify the theoretical analysis that discussed in section 2. Superposition of multi-frequency sinusoidal signal result a radar signals with pulse shapes with a certain period. In this section the simulation is performed with multi-frequency signal of $[f_1, f_1 + \Delta, f_1 + 2\Delta, \ldots f_1 + (N-1)\Delta]$, where $f_1 = 10$ GHz is the lowest frequency, $\Delta = 1$ MHz and N is the number of frequency that used in multi-frequency. 10 GHz is selected Considering to the small displacement resolution scale that need to be achieved. The pulse width is determined by the number of frequencies that used (N). From the radar signal then there is an opportunity to detect a target position through cross-correlation computation.

The figure 3 shows the effect of N on the shape of the radar transmitted waveform. N must be greater than 2 to obtain a pulse signal. Increasing of N cause the pulse signal that reconstructed has narrower pulse width. It can be understood that increasing of N means the increasing the bandwidth. Therefore, the pulse associated to the multi frequency signal has narrower pulse width. The initial position of the target (small displacement) that monitored is then determined by the cross-correlation between the LPF1 output of the reflected signal with the loopback signal. The cross-correlation results of three different target positions are shown in figure 4. According to the result in figure 4, the distance of the target from the radar can be determined and then becomes important information that indicates the location of the small displacement (d_0).

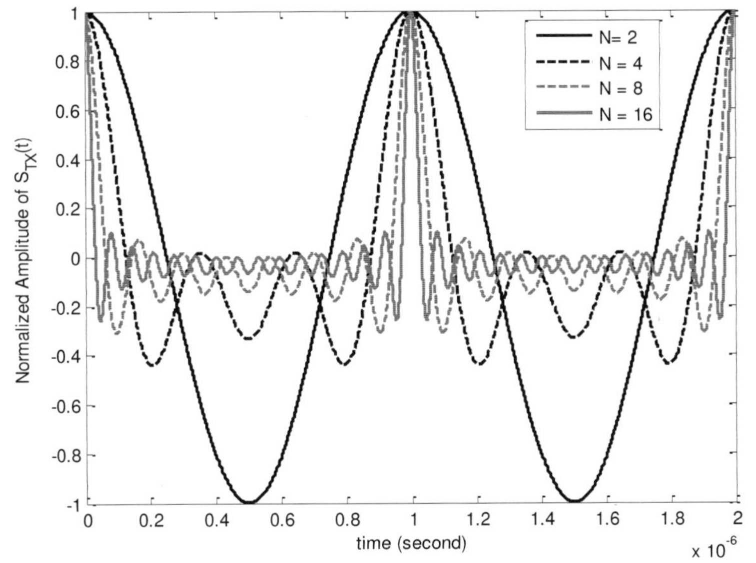

Figure 3. Transmitted signal from multi-frequency CW radar with four different N value.

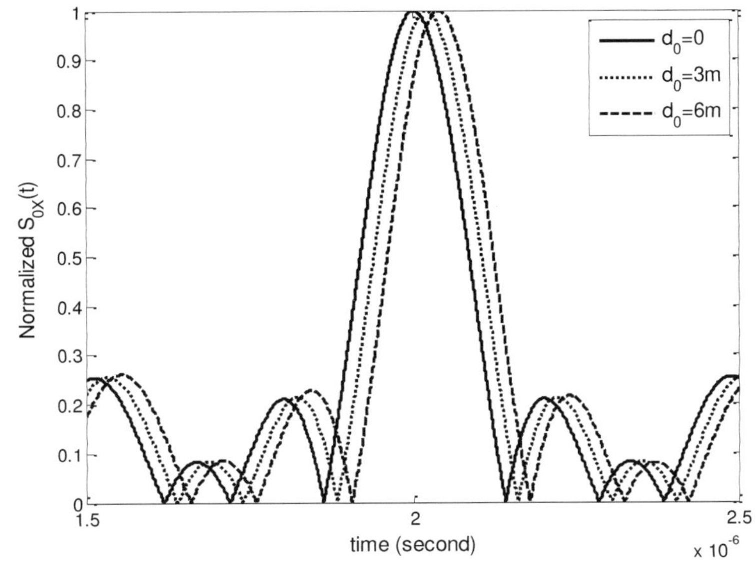

Figure 4. Cross correlation of S_{LPF1} with $S_{loopback}$ that used to determine the target distance (d_0) for three different target locations.

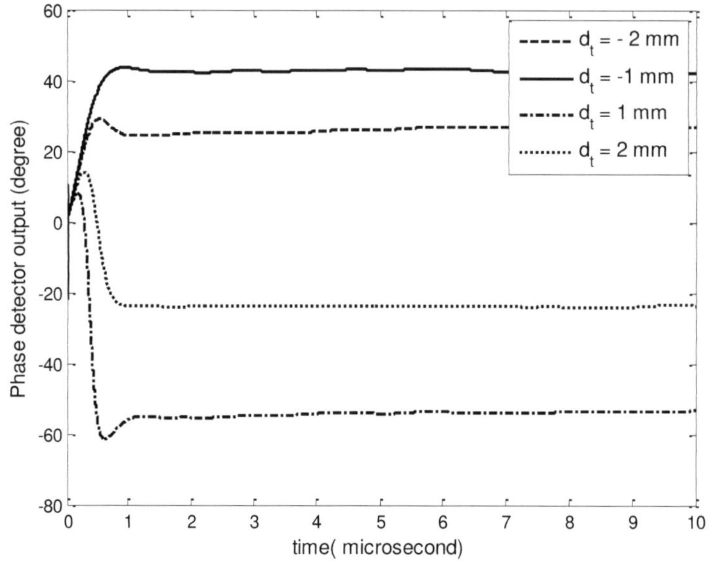

Figure 5. Phase detection result for four different small displacement (d_t) that obtained from IQ demodulation computation.

Consider to the ability of the proposed method in detecting the location of the target, this concept allowing the radar system that proposed having capability for detecting small displacement on multiple targets. However, the accuracy of estimating the target distance from cross-correlation result is limited to the sampling frequency that used to discretize the LPF output. It causes the small displacement at the order of millimetre or centimetre is difficult to detect. In the proposed method, the small displacement (d_t) is determined by computational IQ demodulation as shown in figure 2 against the LPF2 output. The results in figure 5 show the results of small displacement detection for different values at a target with 6 m from the radar. The cross-correlation is used to estimate the location of small displacement and the IQ demodulation is used to estimate the small displacement that occurred at the target. IQ demodulation that is performed in postprocessing side can minimize the change in hardware part.

4. Conclusion

The multi-frequency CW radar was proposed as a method for detecting small displacement. The theoretical and simulation analysis were conducted to investigate the proposed method. The proposed method is a modification of the CW radar by using a multi-frequency generator and two post processing methods. Cross correlation computation is applied to determine the target position whereas IQ demodulation computation is applied to determine the small displacement that occurs. In order to obtain a loopback signal as reference in determining the target distance, a radio frequency switch is added to connect transmitter and receiver section. Switch activation is only done once when starting the measurement and then the loopback data can be stored in memory. Performing the IQ demodulation process in postprocessing part can avoid or minimize the change in hardware part. The results show that the proposed method has ability to detect small displacement at a certain location and possible realized with low bandwidth requirements.

References

[1] A.A Pramudita, Dyonisius Dony and Edwar 2018 Non-Contacting Sensor for small

Displacement and Vibration Monitoring Based on Reflection Coefficient Measurement, (Progress In Electromagnetics Research M, Vol. 7) p. 1–8.
[2] Li C, Chen W, Liu G, Yan R, Xu H and Qi Y. A 2015 Noncontact FMCW Radar Sensor for Displacement Measurement in Structural Health Monitoring (Basel, Switzerland: Sensor, 15 (4)). p.7412-7433, 2015.
[3] K. A. C. de Macedo, F. L. G. Ramos, C. Gaboardi, J. R. Moreira, F. Vissirini and M. S. da Costa, 2017 A Compact Ground-Based Interferometric Radar for Landslide Monitoring: The Xerém Experiment (IEEE Journal of Selected Topics in Applied Earth Observations and Remote Sensing, vol. 10, no. 3).p. 975-986.
[4] Y. S. Lee, P. N. Pathirana, C. L. Steinfort and T. Caelli 2014 Monitoring and Analysis of Respiratory Patterns Using Microwave Doppler Radar (IEEE Journal of Translational Engineering in Health and Medicine, vol. 2) p. 1-12, 2014.
[5] C. H. Hsieh, Y. F. Chiu, Y. H. Shen, T. S. Chu and Y. H. Huang 2016 A UWB Radar Signal Processing Platform for Real-Time Human Respiratory Feature Extraction Based on Four-Segment Linear Waveform Model (IEEE Transactions on Biomedical Circuits and Systems, vol. 10, no. 1) p. 219-230.
[6] Bassem R. Mahafza 2013 *Radar System Analysis and Design*, (CRC Press)
[7] Miquel Testar, Richard Stirling-Gallacher 2011 New super-resolution ranging technique for FMCW radar systems (Proc. SPIE 8188, Millimetre Wave and Terahertz Sensors and Technology IV, 81880K).
[8] A. V. Varavin, G. P. Ermak, A. S. Vasilev, A. P. Yevdokymov and V. V. Kryzhanovskiy 2010 Improvement of range resolution of FMCW autodyne radar (Int.Kharkov Symposium on Physics and Engineering of Microwave, Millimeter and Submillimeter wave, Kharkiv) p. 1-3.
[9] A. Chaudhari, S. Prabhu and R. Pinto 2015 Frequency estimator to improve short range accuracy in FMCW radar (International Conference on Advances in Computing, Communications and Informatics (ICACCI), Kochi) p. 640-644.
[10] B. K. Park, O. Boric-Lubecke and V. M. Lubecke 2007 Arctangent Demodulation with DC Offset Compensation in Quadrature Doppler Radar Receiver Systems (IEEE Transactions on Microwave Theory and Techniques, vol. 55, no. 5) p. 1073-1079.
[11] D. Girbau, A. Lazaro, Á. Ramos and R. Villarino 2012 Remote Sensing of Vital Signs Using a Doppler Radar and Diversity to Overcome Null Detection (IEEE Sensors Journal, vol. 12, no. 3)p. 512-518.

IOP Conf. Series: Journal of Physics: Conf. Series **1195** (2019) 012018 doi:10.1088/1742-6596/1195/1/012018

Sine-Squared pulse approximation using generalized bessel polynomials

Thanavit Anuwongpinit[1], Vanvisa Chutchavong[1], Kanok Janchitrapongvej[2] and Chawalit Benjangkaprasert[1]*

[1]Faculty of Engineering, King Mongkut's Institute of Technology Ladkrabang, Bangkok, Thailand.
[2]Faculty of Science and Technology, South East Bangkok College, Bangkok, Thailand.
*Email: kbchawal@kmitl.ac.th

Abstract. This paper presents the approximation of sine-squared pulse based on the generalized Bessel polynomials. For designing a circuit to synthesize a sine-squared pulse test signal. The generalized Bessel polynomials have more parameter than classical Bessel polynomials that have alpha and beta parameters for adjusting the dominator of the transfer function to approximate the sine-squared pulse that closes to the ideal pulse. The simulation results show that the generalized Bessel polynomial can adjust the approximation response close to the ideal response. The orders of the transfer function are decreased that confirm a better performance than the previous works.

1. Introduction

Sine-squared pulse is impulse response signals that can be applied to many fields of communication system[1]. For example, using to detect deficiency of linear chrominance distortion of colour television transmission system [2]. It can be revealed by the modulated 20T sine-squared pulse test signal. Moreover, an approximation of sine-squared pulse can be used to design matched filters for matching a data between transmitter and receiver[3]. However, the sine-squared pulse cannot synthesis into an electrical circuit. Because it has a format of trigonometry mathematical equation.

In the fields of an analog filter, the filter can be approximated by using mathematical polynomials. It used to synthesize a network function and electrical circuit from a polynomial. Also, the sine-squared pulse signal can be approximated by using some polynomials that have similar some mathematical characteristic. The Bessel filter [4-5] is the all-pole filter having a maximally flat delay and linear phase. The Bessel polynomials have similar features based on the theory of the Bessel Thomson filter. In the previous works, it can be written in form of generalized Bessel polynomials (GBP). The generalized Bessel polynomials have a parameter alpha and beta for adjusting the responses and have some characteristic that similar to the sine-squared pulse signal. For the sine-squared pulse approximation, we introduced the generalized Bessel polynomials as the dominator of the transfer function. The transmission zero pairs are used as a nominator of the transfer function. Unfortunately, the transfer function has high-order degree n of polynomials. It has the complexity to synthesize electrical circuits or networks. Moreover, sine-squared pulse power spectrum in previous works are still far from ideal sine-squared pulse [6-7].

This paper presents a method to solve a problem of power spectrum magnitude and attenuation response of sine-squared pulse by using the generalized Bessel polynomials as a dominator of the

transfer function. It has a parameter to adjust the power spectrum close to the ideal sine-squared pulse. In addition, a polynomials order in a transfer function will be decreased.

2. Generalized Bessel polynomials

The generalized-Bessel polynomials $B_n(s,\alpha,\beta)$ was defined by [8] and the explicit expression of the GBP degree n defined by

$$B_n(s,\alpha,\beta)=\sum_{k=0}^{n}\binom{n}{k}\left(\frac{(n+k+\alpha-2)^{(k)}}{\beta^k}\right)s^{n-k} \tag{1}$$

where $\binom{n}{k}$ is a binomial coefficient and $(z)^k = z(z-1)(z-2)...(z-k+1), z=n+k+\alpha-2$ is the backward factorial function of order k, α and β are real parameters and $\beta \neq 0$. In the previous literature, L. Storch [9] and J. Martinez [10] presented a procedure for Bessel approximation of e^{-s} that the transfer function as

$$h(s)=e^{-st} \tag{2}$$

From Eq. (2), fractional this equation, can be expressed as the exponential form and convert to hyperbolic function as follows

$$h(s)=\frac{\dfrac{1}{\sinh\left(\dfrac{st}{2}\right)}}{\sinh\left(\dfrac{st}{2}\right)\left(1+\coth\left(\dfrac{st}{2}\right)\right)} \tag{3}$$

From Eq. (3), see that the dominator contains the cotangent-hyperbolic function. This characteristic has similar to the characteristic in hyperbolic function of the sine-squared pulse that will describe in the next section.

3. Sine-Squared pulse testing signal

Sine-squared pulse is used for signal testing in baseband communication. Besides, it is a mathematical equation that cannot be used to create matched filter circuits. By equation, sine-squared pulses represent ideal equations. Therefore, it requires specific features of the sine-squared pulse that similar with Bessel polynomials. The ideal of the sine-squared pulse is given as

$$h(t)=\begin{cases}\sin^2\left(\dfrac{\pi t}{2\tau}\right); & |t|\le\tau \\ 0 & |t|\ge\tau\end{cases} \tag{4}$$

From Eq. (4), using the Laplace's Transform and then given as

$$h(s)=\left(\frac{4}{s^2+4}\right)\left(\frac{1}{1+\coth\left(\dfrac{s\tau}{2}\right)}\right) \tag{5}$$

From Eq. (5), see that the result of Laplace's transform of the sin-squared pulse has a hyperbolic mathematical function. It cannot synthesize to a network function or electrical. From section 2, Eq. (3), it has dominator in cotangent-hyperbolic mathematical equations similar characteristic as the ideal sine-squared pulse in Eq. (5). From this similar characteristic, the sine-squared pulse can be approximated and synthesized into network function or electrical circuits. Thus, the GBP has the appropriate characteristic to synthesize the sine-squared pulse signal. From Eq. (4) using Fourier's Transform. Hence, the frequency spectrum of the sine-squared pulse as

$$h(f) = \tau \frac{\sin(2\pi f \tau)}{2\pi f \tau}\left(\frac{1}{1-(2f\tau)^2}\right) \quad (6)$$

Eq. (6) is the equation of the sine-squared pulse frequency spectrum. The magnitude response of the power spectrum of the sine-squared pulse shown in figure 1 and figure 2 shows the attenuation response.

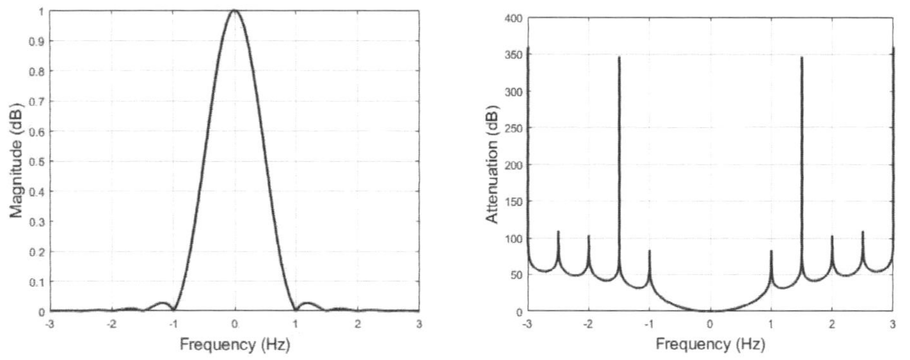

Figure 1. Power spectrum of the ideal sine-squared pulse signal in the frequency domain.

4. Sine-Squared pulse approximation method

For an approximation of sine-squared pulse, the transfer function can be written as polynomials to synthesize the electrical circuit for sine-squared pulse test signal generating. The transfer function has numerator with transmission zero pairs $\pm jk\pi, k = 2, 3, 4, ..., m$. The overall system has minimum stopband attenuation. Therefore, the transfer function $H(s)$ of the filter can be written as Eq. (7)

$$H(s) = \prod_{k=1}^{m/2} (s^2 + \omega_k^2) / \left(\prod_{k=1}^{n} (s - p_k)\right) \quad (7)$$

where m is the degree of the numerator of polynomial, n is the degree of the denominator polynomial, ω_k is the transmission zero pairs $\pm jk\pi$ and p_k is the poles in the left-half of s-plane.

In this paper, the approximation of a sine-squared pulse is considered with the main lobe and first side lobe. The numerator of the transfer function is used two transmissions zero pairs at $\pm j2\pi$ and $\pm j3\pi$. In addition, dominator of the transfer function is used in the GBP due to the Bessel polynomials have a stability with all poles in left-half of s-plane and similar characteristic with sine-squared pulse equation. Therefore, this paper proposes a method to reduce the attenuation by using the GPB as the dominator of the transfer function and varies the parameters alpha and beta in the function.

5. Simulation results

In the simulation results, we define the GBP dominator of transfer function firstly and set two variables of GBP: $\alpha = 2$ and $\beta = 2$ that call as classical Bessel polynomials. The results from previous works have high-order of polynomials. It will cause to complexity synthesize to an electrical circuit. Next, for the approximation of sine-squared pulse, we set the order of GBP are order 3 and set parameters α and β for adjusting the GBP dominator of the transfer function. First, we define $\beta = 1$ and adjust parameter $\alpha = 0.001, 0.1, 1, 10$ and 100. The magnitude response and attenuation response are compared, as shown in figure 2.

IOP Conf. Series: Journal of Physics: Conf. Series **1195** (2019) 012018 doi:10.1088/1742-6596/1195/1/012018

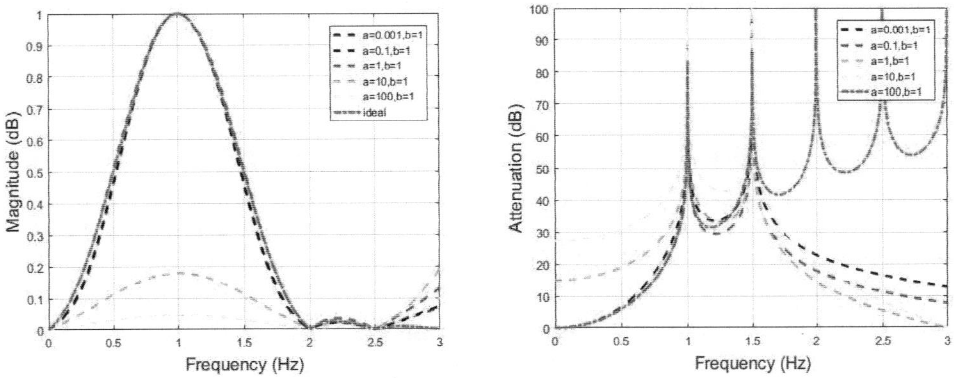

Figure 2. The comparison of magnitude and attenuation response of sine-squared pulse power spectrum with $\beta = 1$ and vary $\alpha = 0.001, 0.1, 1, 10$ and 100.

Figure 3, shows the results of simulation as main lobe are closed with ideal sine-squared pulse when parameter $\beta = 1$ Therefore, we set $\alpha = 0.1$ and varies $\beta = 0.92, 1, 1.2, 1.3$ and 1.5

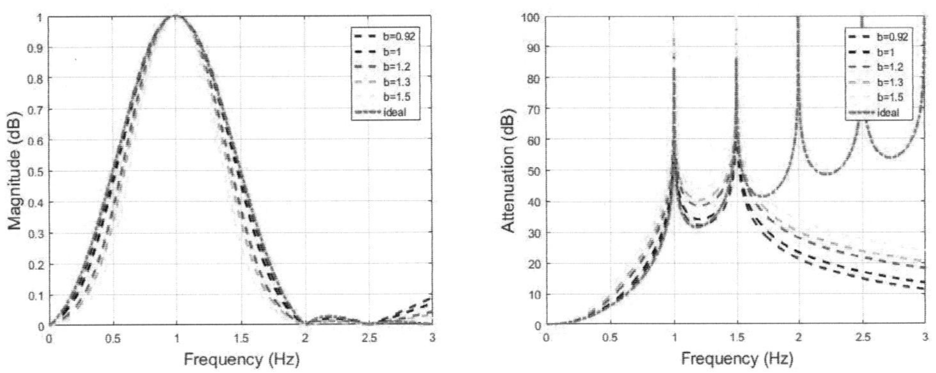

Figure 3. The comparison of magnitude and attenuation response of sine-squared pulse power spectrum with $\alpha = 0.1$ and vary $\beta = 0.92, 1, 1.2, 1.3$ and 1.5.

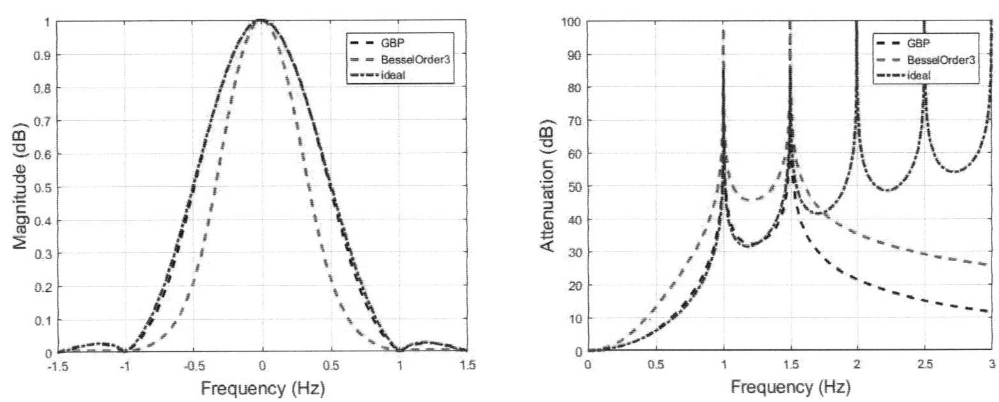

Figure 4. The comparison of the magnitude response of sine-squared pulse power spectrum with classical Bessel polynomials orders 3 and GBP method.

To sum up, the simulation results show the GBP order 3 with $\alpha = 0.1$ and $\beta = 0.95$ that have the power spectrum of sine-squared pulse close to the ideal sine-squared pulse response that summarizes as figure 4.

6. Conclusion

In this paper, sine-squared pulse approximation using the transfer function with generalized Bessel polynomials has been proposed. The generalized Bessel polynomials have a parameter with α and β for adjusting the power spectrum close to ideal sine-squared pulse. The results of the simulations show the generalized Bessel polynomials can control the peak of approximation response by adjusting the α and β parameter. It can decrease the order of the Bessel polynomial to order 3 with generalized Bessel polynomials with $\alpha = 0.1$ and $\beta = 0.95$. Future works, we will consider in the second, third, and next side lobe with the algorithm to define the best α and β parameter and synthesize to the electrical circuit for testing and confirm the performance of approximation method.

References

[1] I. D. Zolotarev and E. M. Lobov, “The research of Gaussian pulse and sine-squared pulse in the implementation of radio-electronic system,” *Proceedings of the 9th International Workshop and Tutorials on Electron Devices and Materials*, pp.140-142, 2008.

[2] V. Chutchavong, O. Sangaroon, C. Benjangkaprasert and K. Janchitrapongvej, "Luminance-Chrominance gain equalizer based on Bernstein polynomials," *IEEJ Transactions on Electronics, Information and Systems,* Vol. 130, pp. 1411-1422, 2010.

[3] S. S. Mneina, G. O. Martens. 2009, “Linear phase matched filter design with causal real symmetric impulse response,” *International Journal of Electronics and Communications*, Vol. 63, pp. 83-91, 2009.

[4] V. Chutchavong, K. Janchitrapongvej, “Gain equalizer approximation by using q-Bessel polynomials,” *In Proceedings of the Asia-Pacific Conference on Communications*, pp. 356-359, 2014.

[5] R. P. Millane and J. L. Eads, “Polynomial approximations to Bessel functions,” *IEEE Transactions on Antennas and Propagation,* Vol. 51, pp. 1398-1400, 2003.

[6] R. Singthongchai, O. Sangaroon and K. Janchitrapongvej, “q-Bessel polynomials for linear phase matched filter,” *In Proceedings of IEEE Region 10 Conference*, 2014.

[7] V. Chutchavong, T. Anuwongpinit, C. Benjangkaprasert and K. Janchitrapongvej, “Approximation of sine-squared pulse with additional transmission zero using Bessel polynomials,” *In Proceedings of the 3rd International Conference on Communication and Information Processing*, 2017.

[8] Eliana X. L. de Andrade, D.K. Dimitrov and A. Sri Ranga, “Characterization of generalized Bessel polynomials in terms of polynomials inequalities,” *Journal of Mathematical analysis and applications*, pp. 538-543, 1998.

[9] L. Storch, “Synthesis of constant time delay ladder networks using Bessel polynomials,” *In Proceedings of IRE*, Vol. 42, pp. 166-167, 1954.

[10] J. Martinez, “Transfer functions of generalized Bessel polynomials,” *IEEE Transactions on Circuits and Systems,* Vol. 24, pp. 325-328, 1977.

IOP Conf. Series: Journal of Physics: Conf. Series **1195** (2019) 012019 doi:10.1088/1742-6596/1195/1/012019

A study of network bandwidth management by using queue tree with per connection queue

C. Smansub, B. Purahong, P. Sithiyopasakul and C. Benjangkaprasert*

Faculty of Engineering, King Mongkut's Institute of Technology Ladkrabang, Bangkok, Thailand

*Corresponding author email: kbchawal@kmitl.ac.th

Abstract. This paper presents a study of network bandwidth management by using the queue tree function with per connection queue function (PCQ). To study a bandwidth management by configuring a network system that uses a Mikrotik router and configures IP number in the network system. WinBox software is used for testing network bandwidth management by download file testing from the internet. The results show a performance of queue tree function and PCQ function for bandwidth management. All users in a network have been received an equal bandwidth of utilization network.

1. Introduction

Nowadays, the internet system and computer networks are necessary for human life. For using a communication, information searching, entertainment, etc., all information on the internet has been continuity developed that includes an audio system, video, and graphics system that causes a big information or data file. From as describes, an organization that uses internet or independent computer networks system. Because a fundamental function of computer networks equipment or tools that use first in-first out queuing technique [1]. It causes information to wait. Benefit for each user bandwidth management can be done using a proxy, it has been providing facilities to manage the bandwidth of one simple queue and PCQ, both facilities have the same function, with different levels of effectiveness seen from the analysis that has been groove carried out from the groove configuration, the effectiveness of both methods [2]. This paper proposes an analysis of computer networks bandwidth management with function queue tree with per connection queue for experimental and analysis of results for network performance.

2. Queue tree function and PCQ

Queue tree is a limit data for certain IP address, subnets, protocols, ports, and another parameter. It is a limit peer-to-peer traffic [3]. Prioritize some packet flows over others. Configure traffic bursts for faster web browsing. Apply different limits based on time. And share available traffic among user equally or depending on the load of the channel. Per connection queue (PCQ) function is one of function that uses for bandwidth management for increasing a performance of the system [3-7]. It divides bandwidth from ISP or related rules from setting menu that divides equally as a user who uses a network at the time.

3. Network design and experiment

3.1 Network connection

This section describes computer networks design that use to test and experiment. The scheme is shown in Fig. 1.

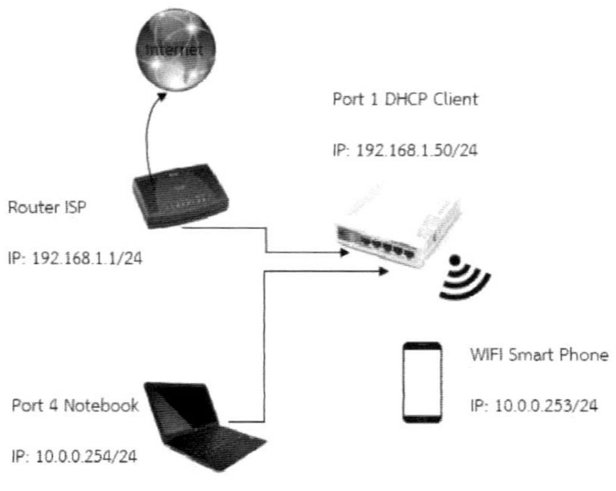

Figure 1. Network design and configuration for this study

Figure 1 presents an infrastructure network design of this system. That shows the connection with LAN and IP configuration of each device in a network to experiment. For the configuration section, Mikrotik router is used [4]. It has a Linux operating system that design for using divide into 2 parts as router operating system and router board. Mikrotik router board hardware is designed by proxy itself by having a variety of series and interface. Router board use router OS for the operating system.

3.2 Queue tree function configuration

In the experiment, we use the WinBox software for configuration network system, function, and experiment. First, we configure a device and router board as shown in Fig. 1, then enable a queue tree function. And then experiment by file download testing size 1 GB from a website that provides a file download testing on the internet. Fig. 2 shows a configuration of queue tree function for maximum bandwidth download. In this experiment, we set a maximum bandwidth of 10 Mbps. In the next step, the configuration of the queue tree to IP 10.0.0.253 and 10.0.0254. For maximum bandwidth upload, we set as same as bandwidth download. A configuration of bandwidth download. In this experiment, we set the maximum bandwidth download is 1 Mbps. A configuration of queue tree with bandwidth upload for IP 10.0.0.253 and we do the same with IP 10.0.0.254.

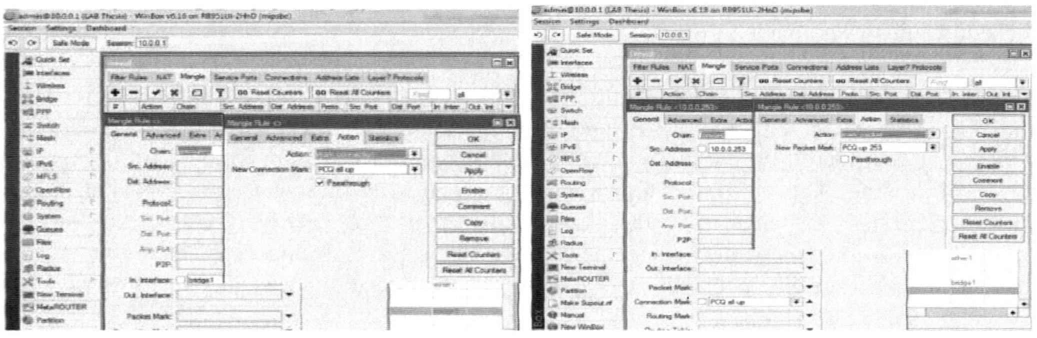

Figure 2. The configuration of mangle rule for out interface and upload configuration to IP 10.0.0.253

3.3 PCQ function setting of queue tree

IOP Conf. Series: Journal of Physics: Conf. Series **1195** (2019) 012019 doi:10.1088/1742-6596/1195/1/012019

In this section, setting a PCQ function configuration of queue tree for maximum bandwidth both download and upload. A configuration of PCQ function in bandwidth download and upload show in Fig. 3.

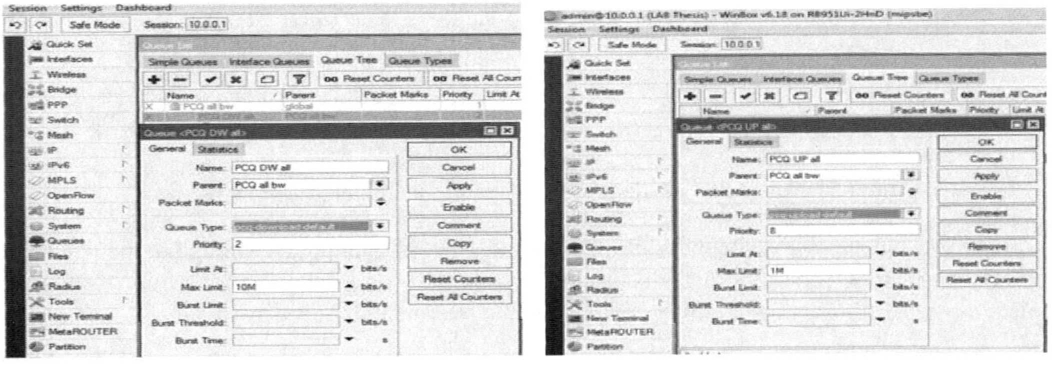

Figure 3. A configuration of queue tree with PCQ function of maximum bandwidth download and upload

4. File download testing and experiment

In this section, we experimented by using a configuration of section 3. And then testing of network bandwidth management by download file in free-ISP download file testing on the internet. Section 4.1 and section 4.2 are the experiment of 1 GB file downloading of the device on the internet by using queue tree function and by using queue tree with PCQ function, respectively.

4.1 Experiment of file downloading by queue tree function

For testing a bandwidth management by queue tree function, we set a maximum limit of download at 10 Mbps and upload 1 Mbps only. This configuration used for setting corresponds with system design in section 3. It not others rules or priority. Fig. 4 shows the windows of file download testing of IP 10.0.0.253 and IP 10.0.0.254 with queue tree function, respectively.

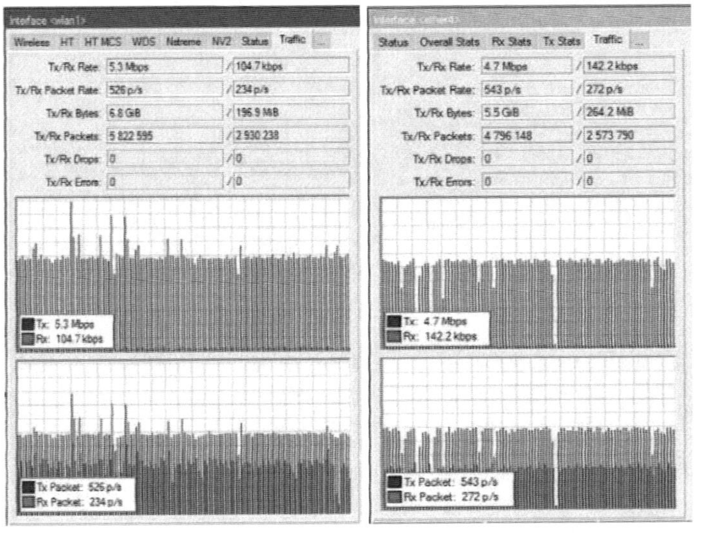

Figure 4. File download testing of IP 10.0.0.253 and IP 10.0.0.254 with queue tree function

The results of file download testing are shown in Fig. 5.

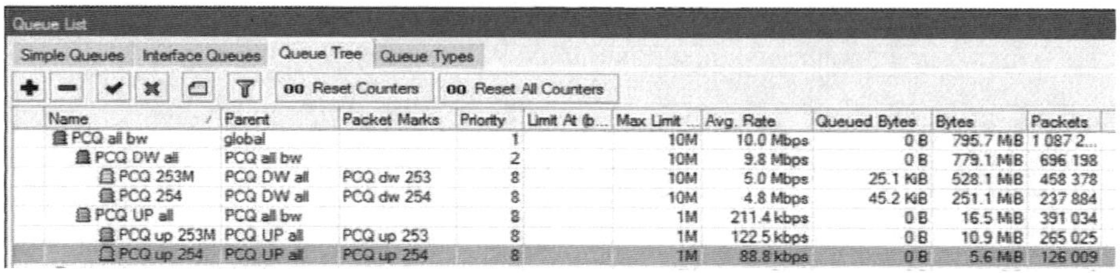

Queue List

Simple Queues | Interface Queues | Queue Tree | Queue Types

oo Reset Counters | oo Reset All Counters

Name	Parent	Packet Marks	Priority	Limit At (b...	Max Limit ...	Avg. Rate	Queued Bytes	Bytes	Packets
PCQ all bw	global		1		10M	10.0 Mbps	0 B	795.7 MiB	1 087 2...
PCQ DW all	PCQ all bw		2		10M	9.8 Mbps	0 B	779.1 MiB	696 198
PCQ 253M	PCQ DW all	PCQ dw 253	8		10M	5.0 Mbps	25.1 KiB	528.1 MiB	458 378
PCQ 254	PCQ DW all	PCQ dw 254	8		10M	4.8 Mbps	45.2 KiB	251.1 MiB	237 884
PCQ UP all	PCQ all bw		8		1M	211.4 kbps	0 B	16.5 MiB	391 034
PCQ up 253M	PCQ UP all	PCQ up 253	8		1M	122.5 kbps	0 B	10.9 MiB	265 025
PCQ up 254	PCQ UP all	PCQ up 254	8		1M	88.8 kbps	0 B	5.6 MiB	126 009

Figure 5. The results of file download testing using queue tree function

From the results, queue tree function has the complexity to config due to an average of the bandwidth of each IP is 1.5 Mbps to 2 Mbps that is continuity. In addition, a period of working time of each IP is completed with a different time 6 minutes and using maximum CPU for processing at 44%.

4.2 Experiment of file downloading by queue tree and PCQ functions

This section enables a PCQ function by changing the checking queue type menu to PCQ-upload default at target upload menu and change to PCQ-download default at target upload menu only. Fig. 6 and Fig. 7 show the windows of file downloading of the device on the internet and the results by queue tree and PCQ functions of IP 10.0.0.253 and IP 10.0.0.254, respectively.

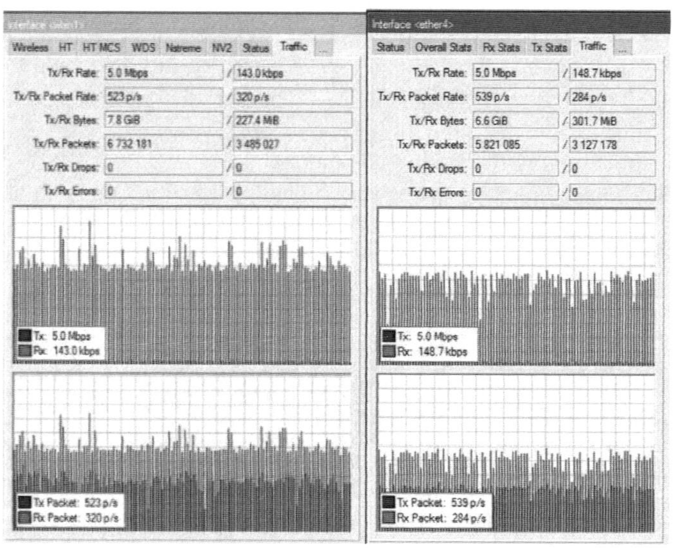

Figure 6. File download testing of IP 10.0.0.253 and IP 10.0.0.254 by queue tree function and PCQ function

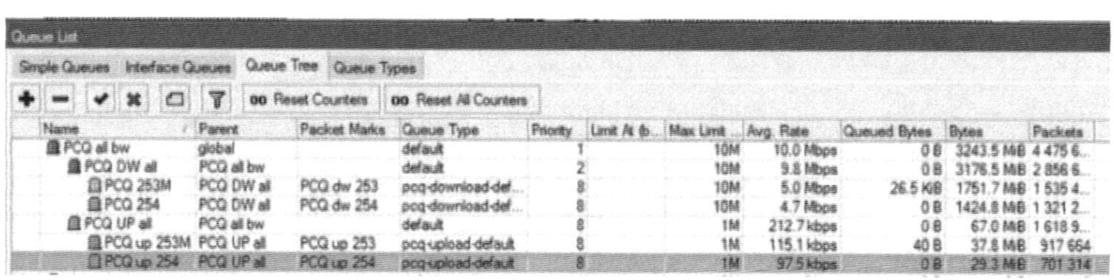

Queue List

Simple Queues | Interface Queues | Queue Tree | Queue Types

oo Reset Counters | oo Reset All Counters

Name	Parent	Packet Marks	Queue Type	Priority	Limit At (b...	Max Limit ...	Avg. Rate	Queued Bytes	Bytes	Packets
PCQ all bw	global		default	1		10M	10.0 Mbps	0 B	3243.5 MiB	4 475 6...
PCQ DW all	PCQ all bw		default	2		10M	9.8 Mbps	0 B	3176.5 MiB	2 856 6...
PCQ 253M	PCQ DW all	PCQ dw 253	pcq-download-def...	8		10M	5.0 Mbps	26.5 KiB	1751.7 MiB	1 535 4...
PCQ 254	PCQ DW all	PCQ dw 254	pcq-download-def...	8		10M	4.7 Mbps	0 B	1424.8 MiB	1 321 2...
PCQ UP all	PCQ all bw		default	8		1M	212.7 kbps	0 B	67.0 MiB	1 618 9...
PCQ up 253M	PCQ UP all	PCQ up 253	pcq-upload-default	8		1M	115.1 kbps	40 B	37.8 MiB	917 664
PCQ up 254	PCQ UP all	PCQ up 254	pcq-upload-default	8		1M	97.5 kbps	0 B	29.3 MiB	701 314

Figure 7. Results of file download testing by using queue tree function and PCQ function

IOP Conf. Series: Journal of Physics: Conf. Series **1195** (2019) 012019 doi:10.1088/1742-6596/1195/1/012019

From the results, queue tree function with PCQ function have an average of the bandwidth of each IP is 5 Mbps that is continuity. In addition, a period of working time of each IP is completed with a different time 6 minutes and using maximum CPU for processing at 46%. Results in section 4.1 and 4.2 present an information to confirm the performance of queue tree function with PCQ function that a queue tree with PCQ has an average bandwidth of each IP is 5 Mbps that more than using queue tree function only.

5. Conclusion

This paper presents a studying of network bandwidth management by using queue tree functions with PCQ. We configurated a device and used Mikrotik router and set IP in the network system and then using WinBox software for testing a network performance of the function. The results show a performance of queue tree function. It may be complex to use due to it has more detail about the configuration follow as rules or limits. PCQ function has more average bandwidth than the queue tree function. Because PCQ not calculate the bandwidth of each IP in download and upload from ISP. PCQ calculated bandwidth from ISP and user in a network. All users in a network have been received an equal bandwidth of utilization network.

References

[1] R. Akbar and R. Indrajit, "Analysis of effectiveness of using simple queue with per connection queue (PCQ) in the bandwidth management (A case study at the academy of information management and computer mataram (AMIKOM) mataram)", Journal of Theoretical and Applied information Technology, vol. 83, pp. 319-326, 2016.

[2] M. M. G. Rashed and M. Kabir, "A comparative study of different queueing techniques in VOIP, video conferencing and file transfer", Daffodil International Journal of science and technology, vol 5, pp. 37-47, 2010.

[3] A. S. Tanenbaum and D. J. Wetherall, "Computer Networks", 5th ed., Prentice Hall, 2011.

[4] D. A. Asmoro, H. Susilawati, A. Wisnu and W. Nugraha, "Implementasi HTB (Hierarchical Token Bucket) untuk manajemen bandwidth pada router Internet Di Unniversitas jenderal soediman", Journal of Techno, vol.12, no. 2, October 2011.

[5] A. Syaifuddin, M. Yunus and R. Sundari, "Perbandingan metode simple queues dan queues tree untuk optimasi manajemen bandwidth jaringan komputer di stmik ppkia pradnya paramita malang", Jurnal Teknologi Informasi, vol.4, pp. 60-74, 2013.

[6] A. Haimi, "The influence of bandwidth restriction to network performance", National Conference of System and Informatics, Bali, pp. 6, 2011.

[7] R. I. Perwira. and F. Liantoni., "Queue tree implementation for bandwidth management in modern campus network architecture", KINETIK, vol. 3, no. 1, pp. 17-26, February 2018.

IOP Conf. Series: Journal of Physics: Conf. Series **1195** (2019) 012020 doi:10.1088/1742-6596/1195/1/012020

Ensemble learning based Architecture Vulnerability Factor calculation using partial feature set in processors

Jiabin Wang[1], Jiajia Jiao[1]*, Yuzhuo Fu[2]

[1] Shanghai Maritime University, Shanghai, China
[2] Shanghai Jiao Tong University, Shanghai, China

Email: jiaojiajia@shmtu.edu.cn

Abstract. With the scaling technology, soft error induced bit upsets are increasingly threatening the processor reliability. Processor designers require effective tools or methodologies to estimate the often-used metric Architectural Vulnerability Factor (AVF). This paper presents an ensemble learning based AVF calculation methodology for fast reliability assessment. Instead of the entire feature set, only partial non-critical attributes are selected to build the predictive model so that many performance counters can be removed or shut down for saving memory space and power consumption. Millions of data collected from a cycle-accurate simulator sim-SODA, are trained by the latest learning methods in Tensorflow. The SPEC2000 results demonstrate the instanced ensemble learning-random forest and Ada-boost perform nearly perfect accuracy, better than linear regression, and neural network.

1. Introduction

The scaling technology makes soft error dominate in the computer reliability design. Due to the increasing error rate and complex patterns of multi-cell upsets, the more and more concern on soft error have received from the academic and industry. The designers of reliable processors should have effective reliability assessment methods or tools to support the dynamic configurations or early design guide.

Architectural Vulnerability Factor (AVF), as the often used metric, represents the probability of one bits upset occurring resulting in a wrong output in the defined observation point. It is usually evaluated by the fast Architecture Correct Execution (ACE) analysis [1] or accurate fault injection [2]. The former can make good use of the component utilization for the lower bound of AVF while the later adopts lots of time consuming simulations for characterizing the soft error impacts on processors. Few works also try some analytical models for fast evaluation of soft error influences [3]. To further meet the demands of online configurations for different benchmarks or architectures, the paper work aims at the machine learning methods applied into fast and accurate AVF predication.

2. Related work

Dynamic prediction of architectural vulnerability from micro-architectural state was proposed by identifying strong correlations between structural AVF values and a small set of processor metrics [4]. AVF stressmark was generated by an automated methodology for bounding the worst-case vulnerability to soft errors in [5]. The online estimation of architectural vulnerability factor for soft errors was implemented [6]. Unlike the previous work, the machine learning method of boosted

regression tree was used to predict micro-architectural vulnerabilities in [7]. A linear regression was used for predicting the detectable and unrecoverable errors (DUE_{AVF}) and silent data corruptions (SDC_{AVF}) [8]. Machine learning technique involved for off-line training of an application with representative inputs, and on-line detection using the model, applied even to a different dataset [9]. Machine learning based generic run-time method with low area and power overhead was presented to predict the soft-error vulnerability of on-chip memory arrays as well as logic cores[10]. The existing works with the chosen machine learning method perform well from the various experiments or verification. However, there is lack of comprehensive comparison to point the more appropriate learning approach in the AVF prediction.

As a complementary work, the paper compare the classical or hot learning methods (linear regression, regression tree, as well as neural network), we recommend an ensemble based cost-effective methodology to predict AVF using the partial performance metrics.

3. Proposed ensemble learning based AVF assessment methodology

3.1. Observation and analysis

A lot performance metrics can be collected as the attribute features. For example, the ACE analysis platform [11] provides about 190 metrics, e.g., sim_num_loads, simu_num_stores, sim_num_insns. To remove some invalid features, the principal component analysis (PCA) [12] is used in figure 1 to draw 10 features impacts on AVF. The distinct linear trend can be found for each SPEC2000 benchmark. However, the mixed results of all benchmarks show half linear region and half nonlinear region. Based on the observation, we can refer the linear regression can work in AVF prediction but not perfect due to the nonlinear factors.

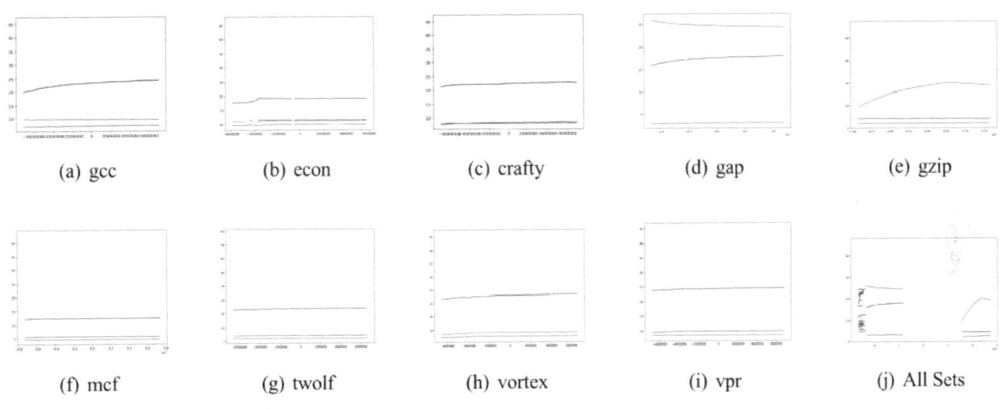

Figure 1. AVF estimation based on principal component analysis.

3.2. Case study of ensemble learning for AVF prediction

Unlike interpretable linear regression, as a white box to represent the correlation of output and features clearly and directly, some linear based extensions are also attractive in the data analysis filed. Backward propagation neural network can be considered a special linear extension with nonlinear activation function (e.g.Sigmoid, ReLU, tanh). It works like a black box to handle the potential nonlinear factors. Ensemble learning like random forest and Xg-boost uses multiple week learning systems instead of a single strong learning system [13][14]. It seems a grey box, a trade-off between linear regression and neural network. It is referred that the ensemble learning works well in AVF prediction due to the grey box based compromise.

Random forest and Xg-boost are the represents of ensemble learning. The case study of random-forest for better understanding the ensemble learning is depicted in figure 2. The main idea of random

forest is to use regression tree for training each randomly selected data segment and sum up all of the weak training results as the final output.

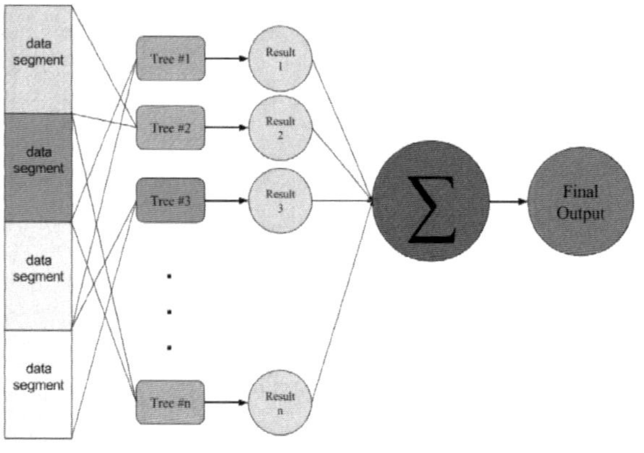

Figure 2. Random forest: case study of ensemble learning.

4. Simulation and analysis

4.1. Simulation framework and configuration

To apply the machine learning methods into the AVF estimation, we build a complete simulation framework in figure 3. The overall framework combines the sim-SODA and tensorflow[15] together. Firstly, the millions of dataset are generated from SPEC2000 benchmarks execution in sim-SODA. The sampling interval is 1000 instructions for each benchmark. Secondly, the dataset are processed by data integration, feature selection and data clean respectively. Then, the clean data used for training and testing for different machine learning methods in tensorflow. Finally, two often-used metrics MAE and R2 are used for evaluation to decide the more appropriate approach.

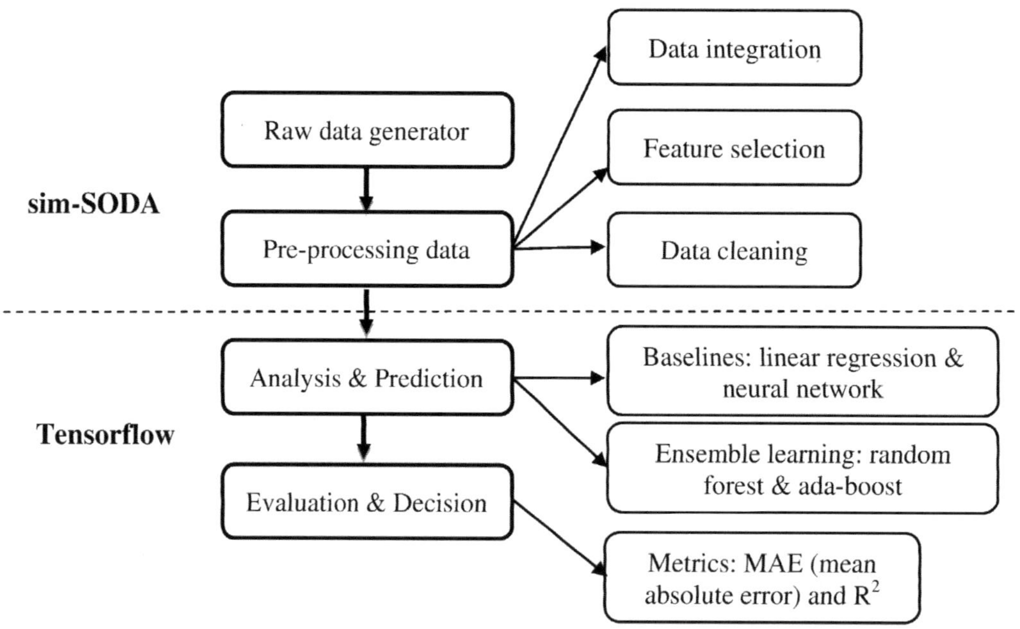

Figure 3. Overall simulation framework of sim-SODA and tensorflow.

4.2. Results and analysis

The IQAVF (AVF for instruction queue) prediction can be evaluated by mean absolute error (MAE) and fitting excellence R^2. Local 30 features about instruction and data TLB are used for training. The evaluation results of MAE and R2 for different learning methods comparison are listed in figure 4 and Table1 respectively.

From the MAE results in figure 4, we can find out ensemble learning has nearly perfect mean absolute error. Instead, linear regression and neural network performs not well. Neural network is even worse than linear regression because of the linear trend in AVF estimation of subsection 3.1. This phenomenon is consistent with the theoretical analysis of learning ideas in subsection 3.2. Random forest and Ada-boost can takes full advantage of the multiple small week learning systems in the case of some local features training.

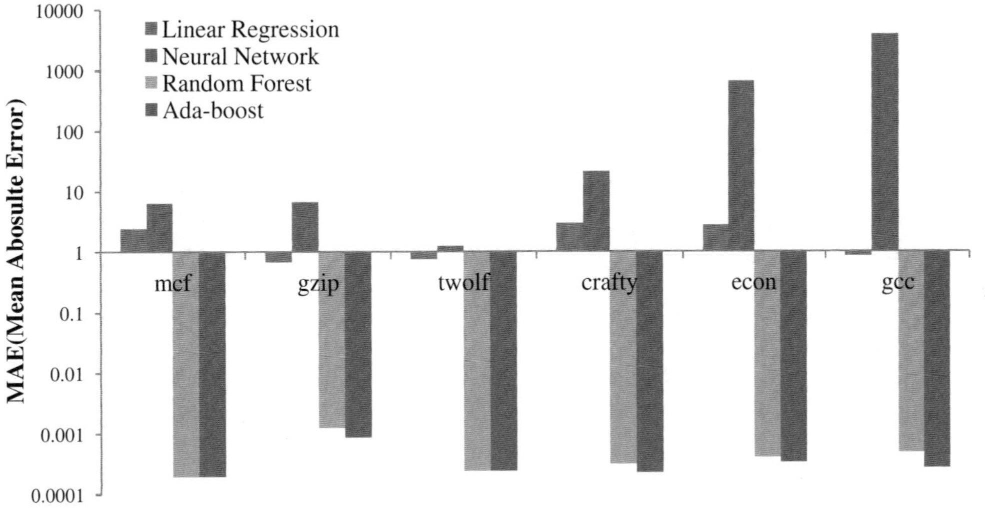

Figure 4. MAE evaluation results comparison.

From the perspective of fitting excellence R^2, the same conclusion can be drawn from Table1. Linear regression and neural network have the abnormal results beyond the legal range [0,1]. It is reasonable for their powerlessness of using local features to predict the global AVF value. The recommended ensemble learning achieve the dominate win for a serial values of 0.999.

Table 1. R^2 evaluation results comparison.

Learning method / Benchmark	Linear Regression	Neural Network	Random Forest	Ada-boost
mcf	1.02E+08	-68186.6	0.999896	0.999996
gzip	0.835267	-5.8744	0.999999	0.999999
twolf	15061664	-11115.6	0.999346	0.999278
crafty	13254025	-200028	0.999898	0.999915
econ	208803.8	-2.2E+07	0.999979	0.999975
gcc	955554.7	-1.7E+10	0.990333	0.999673

In all, the above results demonstrate that the ensemble learning method is cost-effective and recommended for reliable processor design.

IOP Conf. Series: Journal of Physics: Conf. Series **1195** (2019) 012020 doi:10.1088/1742-6596/1195/1/012020

5. Conclusion

The paper presents an ensemble based cost-effective methodology to predict AVF using the partial performance metrics. The overall simulation framework of sim-SODA and tensorflow is built for comprehensive comparison. The results of both MAE and R2 show that the ensemble learning uses the selective 30 local features for 99.9% AVF prediction accuracy, better than linear regression and neural network.

References

[1]. Mukherjee S S, Weaver C, Emer J, et al. A systematic methodology to compute the architectural vulnerability factors for a high-performance microprocessor[C]// Microarchitecture, 36th Annual IEEE/ACM International Symposium on. IEEE, 2003: 29-40.

[2]. Leveugle R, Calvez A, Maistri P, et al. Statistical fault injection: Quantified error and confidence[C]//Proceedings of the Conference on Design, Automation and Test in Europe. European Design and Automation Association, 2009: 502-506.

[3]. Nair A A, Eyerman S, Chen J, et al. Mechanistic modeling of architectural vulnerability factor[J]. ACM Transactions on Computer Systems (TOCS), 2015, 32(4): 11.

[4]. Walcott K R, Humphreys G, Gurumurthi S. Dynamic prediction of architectural vulnerability from microarchitectural state[J]. ACM SIGARCH Computer Architecture News, 2007, 35(2): 516-527.

[5]. Nair A A, John L K, Eeckhout L. AVF stressmark: Towards an automated methodology for bounding the worst-case vulnerability to soft errors[C]//Proceedings of the 2010 43rd Annual IEEE/ACM International Symposium on Microarchitecture. IEEE Computer Society, 2010: 125-136.

[6]. Li X, Adve S V, Bose P, et al. Online estimation of architectural vulnerability factor for soft errors[C]//ACM SIGARCH Computer Architecture News. IEEE Computer Society, 2008, 36(3): 341-352.

[7]. Li B, Duan L, Peng L. Efficient microarchitectural vulnerabilities prediction using boosted regression trees and patient rule inductions[J]. IEEE Transactions on Computers, 2010, 59(5): 593.

[8]. Wibowo B, Agrawal A, Tuck J. Characterizing the impact of soft errors across microarchitectural structures and implications for predictability[C]//Workload Characterization (IISWC), 2017 IEEE International Symposium on. IEEE, 2017: 250-260.

[9]. Liu J, Agrawal G. Soft Error Detection for Iterative Applications Using Offline Training[C]//High Performance Computing (HiPC), 2016 IEEE 23rd International Conference on. IEEE, 2016: 2-11.

[10]. Vijayan A, Kiamehr S, Ebrahimi M, et al. Online Soft-Error Vulnerability Estimation for Memory Arrays and Logic Cores[J]. IEEE Transactions on Computer-Aided Design of Integrated Circuits and Systems, 2018, 37(2): 499-511.

[11]. Fu X, Li T, Fortes J. Sim-SODA: A unified framework for architectural level software reliability analysis[C]//Workshop on modeling, benchmarking and simulation. 2006.

[12]. Wold S, Esbensen K, Geladi P. Principal component analysis[J]. Chemometrics and Intelligent Laboratory Systems, 1987, 2(1):37 – 52. Proceedings of the Multivariate Statistical Workshop for Geologists and Geo- chemists.

[13]. Breiman L. Random forests[J]. MachineLearning,2001, 45(1):5–32.

[14]. Chen T, Guestrin C. Xgboost: A scalable tree boosting system. Proceedings of the 22Nd ACM SIGKDD International Conference on Knowledge Discovery and DataMining, NewYork, NY, USA: ACM, 2016. 785– 794.

[15]. Abadi M, Barham P, Chen J, et al. Tensorflow: a system for large-scale machine learning[C]//OSDI. 2016, 16: 265-283.

Acknowledgments

This work is supported by National Natural Science Foundation, numbered 61502298, 61472244 and Innovation Program of Shanghai Maritime University.

IOP Conf. Series: Journal of Physics: Conf. Series **1195** (2019) 012021 doi:10.1088/1742-6596/1195/1/012021

Integration of multilayered context-aware control system for ubiquitous computing environment

Khamla Non Alinsavath[1*], Lukito Edi Nugroho[2] and Widyawan[3]

[1, 2, 3] Department of Electrical Engineering and Information Technology, Faculty of Engineering, Universitas Gadjah Mada, Yogyakarta 55281, Indonesia.

Email: khamla@mail.ugm.ac.id[1], lukito@ugm.ac.id[2], widyawan@ugm.ac.id[3]

Abstract: In response to recent demand, context-aware application performs significantly in the field of pervasive computing and smart environment. Many approaches have been proposed to provide functionalities and implementation using context information to observe the environment. Developing context-awareness system becomes more essential issue in order to support into the real situation and system deals with many components and complexity. There are many hardware with different platforms that make them have difficulty to communicate to each other with different layers processing in the system of context such as physical, middleware and application layers. Services in each layer need to adapt the way they behave according to the current context information. This research introduces the development of context-aware system and middleware developments that support the creation of smart application with the purpose of providing the framework of the system and deploy middleware as being the bridge to integrate physical layer and application layer seamlessly to support monitoring ubiquity environment. The system controls multi sensors to monitor the real environment from raw data through prototype platforms for managing, sensing, computing and user interacting as the outcome of the system's performance. Implementation and experiment have been performed where user moves in / moves out from the sensor range. The proposed middleware is able to handle the complicated system and fulfill its requirement such as heterogeneity, scalability and mobility, the system and complicated components are integrated to be a completed system that operate, interpret and representation context entities to an end users. Having this control system makes a lot of benefits such as real time information access, become intermediary between layers and support to control energy system with the purpose of reducing the energy consumption, rapid communication to the users.

1. Introduction

Context awareness [1] is an important part of pervasive computing field to monitor and control human behavior and real environment. A system can lead to the development of a rich knowledge-based that will not only be useful for society, but also serve as an essential input for a system's processing. The definition of the context is from Schilit Adams [2] which introduce the concept of context is related to the location, nearby person, hosts or objects as well as changes of them over the time context. Context [3] is defined as situational information of entities (place, object, or people) [4], the system has to acquire context correctly and provide adequate service behaviors regarding to provided context. Context is divided into three categories: computing context, user context, physical context. In order to achieve better understanding more about context, Chen and Kots [5] add time context such as time of a

day, week, month and season of the they year. Time information like a context's history that adds to this definition.

Context awareness is the ability of a program or computing device to send and act upon the information in its environment such as location, time and user identities [6][7]. In the real environment, context changes dynamically [8] and the applications have to adapt their behavior, therefore context-aware system need to manage its context information effectively and satisfy qualities of context information required by context-aware applications [9]. The real-time capability of new human behavior monitoring solutions can provide mission critical artificial intelligence [10]. Hence computing infrastructure for understanding human movement processing is increasingly important, especially when merging this computing infrastructure and technology in order to obtain more advantage and efficiency while processing data with innovation, transformative methodology and successfully solutions. Context awareness [11] [12] also can be broadly classified into the following areas: self-contains and infrastructure based context awareness, context information is important to apply to the system and situation awareness [13] such as location, identity, time and activity that follow with these questions of 5Ws: who, what, then, where and why, which is minimum necessary set of context information in the pervasive computing environment.

This research aims to develop context-aware application to deal with existing issue, to understand user events and integrate the system into seamless application by middleware that can support and being intermediary between physical and application layers without obstacle, it is frequently protected by more efficient monitoring techniques. Middleware can be a solution for a complicated system to access data easily and comfortably, and also to fulfill the requirement of context-aware middleware such as heterogeneity, mobility and ease of deployment. On the other hand, context-aware application has been proposed to implement together with middleware and become application program interface to read data from raw sensors through middleware and to monitor the environment conditions.

The remainder of this paper is organized as follows: section II provides background and related works on applying context-awareness scenarios and ubiquitous computing. Section III introduces multilayered context-aware system architecture including hardware, middleware, and application frameworks. In the section IV, the performance of the experiment, result and discussion are presented. Section V concludes the entire work of this paper and future work for the paper.

2. Related Works

The context-aware system [14] as the crucial component in Ubiquitous Computing [15], perceives context changes and adapts accordingly. It provides an important way to improve intelligence system, and comes up with a convenient and efficient interaction method between the intelligence system and the surrounding environment by middleware. The context-aware system has a general architecture divided into three tiers: context-aware application, context-aware service, and context sensors. Many researchers have proposed different context aware middleware schemes to overcome different problems. To find a simple, fast and qualified algorithm for middleware application based on context awareness is a scenario widely studied by a lot of researchers. Limon et.al [16] presented a context aware middleware architecture which provides support for both pervasive and ad hoc computing. The middleware is structured in terms of a number of component frameworks, for example, the middleware platform provides support for dissemination of events in a timely fashion. Mechanism for context awareness and intelligent decision-making are also provided. Lastly a flexible framework is offered for enabling the use of a number of devices service discovery protocols. Bai et.al [17] presented the middleware to support the application development of context awareness under the new-type networks environment represented by wireless sensor network. It applied the updated service-oriented and light-weight structure with excellent expansibility and efficiency in the running process. Additionally, the architecture of middleware mainly contains five layers and cross two layer modules: hardware abstract, service registry, context model, reasoning and application representation layers; energy management module and security module. During the process of context awareness, the new method of awareness synchronization is designed to ensure sensitivity to context switch.

IOP Conf. Series: Journal of Physics: Conf. Series **1195** (2019) 012021 doi:10.1088/1742-6596/1195/1/012021

3. Multilayered Context-Aware System Architecture and its application

3.1. Multilayered Context-Aware System Design

We present the design and architecture of the context-aware system which serves the platform of desktop applications. It provides a mean to store and retrieve contextual information, and also facilitates providing relevant services to the application so that the context information can be used more effectively. The system communicates with the database and other service for gathering context information and then present it to the client applications. This proposed application is also designed to have advanced context usage functionalities such as learning from human behaviors, weather and light intensity. Additionally, entities such as people, objects and location would themselves serve as contextual information sources. The system is designed to aid a human decision maker and help people that interact with a user interface, and the designed architecture has essential features of context-aware systems namely managing, controlling, filtering data and intermediary between layers.

Most of the proposed systems are user centric and the context of a situation is defined only with respect to the entities involved in the events, activities, properties and relationships between them. To tackle these drawbacks, we have developed the context-aware middleware that is being designed to support the requirements of the middleware in the pervasive environment such as the proposed system can support for heterogeneity, scalability, mobility and so on. It supports integration of different application and collection and storage information from heterogeneous services.

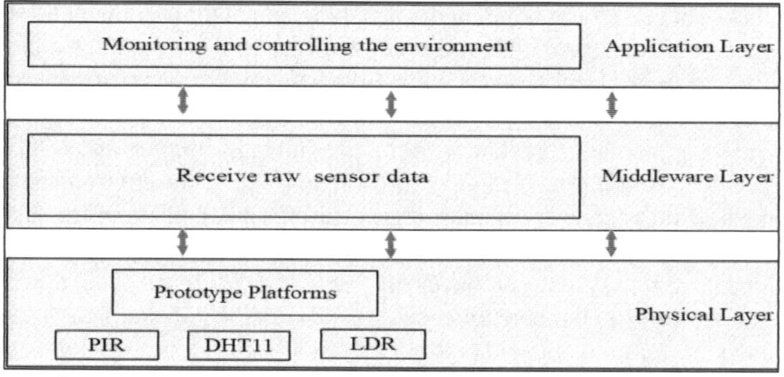

Figure 1: Context-aware System architecture overview

Figure 1 shows that the context-aware system consists of three layers such as the application layer, middleware layer and physical layer. The details of each layer are listed below:

The physical layer consists of four kind of sensors namely Passive Infrared sensors, light-dependent resistor and Temperature and Humidity sensors, they are used to capture and communicate with the real environment through prototype platforms. The middleware layer performs as a bridge between physical and application layers and being integration point for the system seamlessly. After receiving data from the hardware then the middleware will manage and control the data and collect data from the physical layer and data will be recorded to the database. The application layer is the top layer of the proposed system that is used to communicate with the users by monitoring human activity, controlling the system and providing data when users request data through the system from sensors.

3.2. Proposed Context-aware middleware for the system

The middleware provides an abstraction and reliable services for common operations and would simplify the development of context aware applications. It would be accommodated a variety to hardware, operating systems and programming languages. The middleware would also allow us to compose complexity of the systems based on the interaction between users and context aware applications. More significantly, it supports for complex concern such as contextual information in the hardware layer as well as raw sensors. It is also defined a common model context that to ensure the different application in the smart environment.

IOP Conf. Series: Journal of Physics: Conf. Series **1195** (2019) 012021 doi:10.1088/1742-6596/1195/1/012021

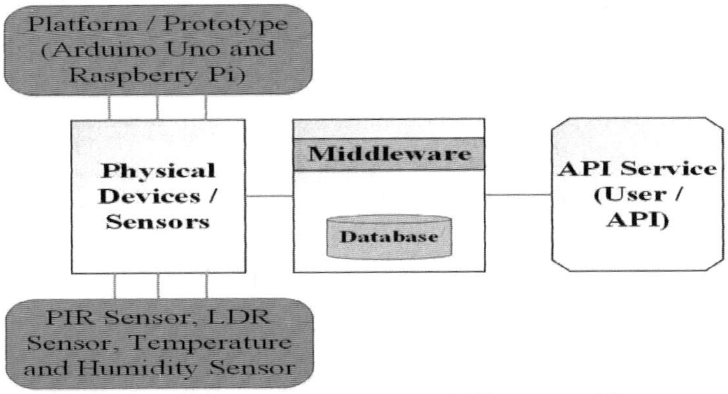

Figure 2: Proposed Context-aware middleware architecture

The proposed middleware specifies certain requirements for middleware's in the smart environment, for which terms of requirement are as follows:

1. The proposed middleware should support collection of context information from heterogeneous sensors and services and interaction of appropriate context information from real conditions.
2. It should support the mobility of the system uses and efficiency interface to monitor the system.
3. The application layer can easily access data from the middleware layer, it simply make a request to the middleware and then middleware will manage then deliver the requested data to the application.
4. It should handle complexity of the system and different contexts between application and services.

3.3. Context-aware service and application processing

The context-aware service diagram that illustrated in figure 3 collects information about the current environment which receives from the real sensors and sends received data such as human movement, temperature, light intensity and humidity to the system to filter and analyze as real time services. The information can be updated by the user with the application and raw sensors communicate to each other through the proposed middleware.

Context–aware services adapt to the environment of the user automatically and the system allows for interactions between user and application by different kinds of sensors that control the light bulb and air-conditioner, along with human activities such as moving in - out the sensor range and light intensity.

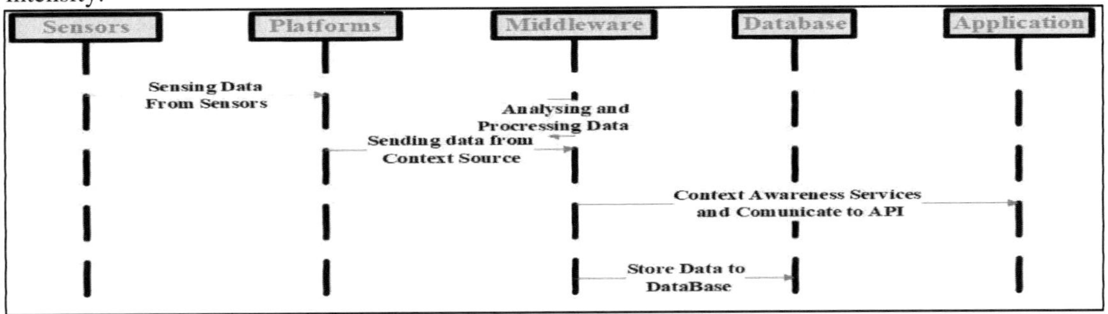

Figure 3: Context-Aware service diagram

The figure 3 presents the processing of the application which works properly according to all components. All of the data from the sensors captured from real conditions are sent to middleware through the TCP/IP protocol and the Serial protocol, after that the collected data are sent from middleware to the database.

4. Result and Discussion

4.1. *Experimental Design*

The experiment starts from the physical layer, then moves through the middleware layer and finishes with application layer. We have implemented an equipment setup that consists of many devices such as sensors, prototype platforms, and personal computer which are connected by serial and Ethernet cables. The whole system is illustrated in the figure 4.

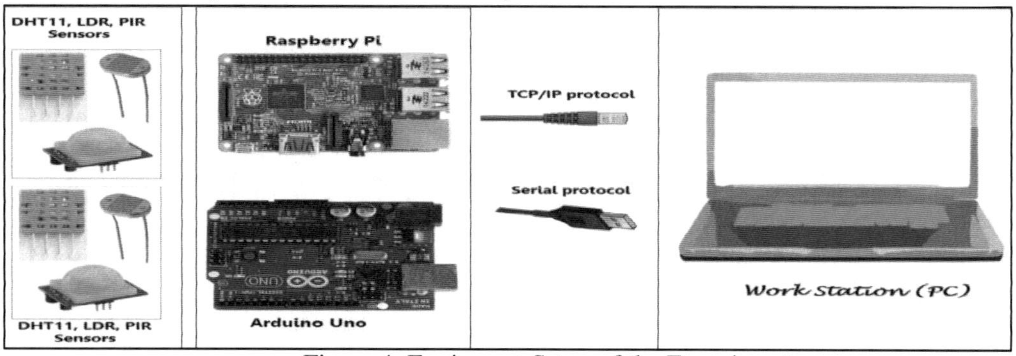

Figure 4: Equipment Setup of the Experiment

The result of the proposed system can appropriately capture the conditions and situation. If situation detected is mean that there is someone moving within the sensor range. As a result, the LED (light-emitting diode) will turn on. The LED will stay on as long as the people move around, and after a period where no movement is detected, the LED will turn off. The situation is depended on data from middleware layer which come from real detection of the environment. The middleware layer deals with two different systems and eight sensors (four sensors for Arduino platform, four sensors for Raspberry platform) and collects all of the real data from the physical layer. Moreover the middleware is responsible for arranging the required data when users request. Sensors provide data about events in the real situation and classify the situation into context such as time context, location context and presence context after collecting all of data then transferring them to the middleware layer.

4.2. *Middleware layer implementation and integration point*

To evaluate the performance of the proposed middleware system, the requirements of the middleware that we need to fulfill are essential to implementing and developing a better system. This middleware layer becomes intermediary between physical and application layers, also it integrates whole system to be seamlessly application with the completion of service and provide raw data to the users through this middleware layer. After the function is completed and able to monitor the system from the middleware layer, all of raw data from the sensor are stored in the database (table 1). The latest data received from multi sensors will be stored for one month and then will be replace new data automatically. The middleware layer will receive all of the services from the hardware layer that work to completely success and cover the complexity of physical services. All of collected data are then ready to be sent to the application layer. In order to connect with other processes, we need to have other packages or services to communicate with sensors by using protocols such as the Serial protocol which is used to communicate with Arduino and multi-sensors, or the TCP/IP protocol, which is used to communicate with Raspberry and multi-sensors.

Table 1: Database for collecting data from monitoring environment

ID	Arduino_ LDR	Arduino_ Humid	Arduino_ Temp	Arduino_ PIR	Raspberry_ LDR	Raspberry_ Humid	Raspberry_ Temp	Raspberry_ PIR
876	26	47	25	0	97	49	25	0
877	25	46	17	0	93	49	25	0
878	24	46	25	0	91	49	25	0
879	25	46	25	0	89	49	25	0
890	25	46	25	0	96	49	25	0
891	26	46	50	0	95	49	25	0

IOP Conf. Series: Journal of Physics: Conf. Series **1195** (2019) 012021 doi:10.1088/1742-6596/1195/1/012021

The table 1 shows the latest data received from the multi-sensors that will stored for one month. The data consists of 2 different systems and 6 values of data from the sensors such as LDR sensor, Humidity and Temperature sensors and PIR sensors.

4.3. Application Layer implementation and result

We have implemented the application layer called the context-aware application to monitor and control the whole system. There are two kind of interfaces that we have implemented such as package data that shows one hundred data from database that bring from middleware which illustrated in the table 3 and single data that shows the latest data from real environment in the table 4. The application will directly communicate with the middleware, after which the middleware will handle the physical layer and obtain data from sensors to illustrate in the provided database and prepare to show in the API interface. The proposed middleware is able to deal with the complexity of the system in the hardware components that makes it easier for users to use and operate the completed system. Middleware hides the complexity of the hardware layer so that the application could access the middleware anytime when user inquires.

Table 2: Assigned data to illustrate on application layer

```
jsonObject.put("raspberry_LDR", dt[0]);
jsonObject.put("raspberry_humidity", dt[1]);
jsonObject.put("raspberry_temperature", dt[2]);
jsonObject.put("raspberry_PIR", dt[3]);
jsonObject.put("arduino_LDR", dt[4]);
jsonObject.put("arduino_humidity", dt[5]);
jsonObject.put("arduino_temperature", dt[6]);
jsonObject.put("arduino_PIR", dt[7])
String result = jsonObject.toString();
```

Users interact with the application layer directly from the application through web browser. The application layer implements a communicating component and the availability of all communication between applications requires cooperation managed by the system and application layer. The end-user processes are supported by this layer, the quality of service is identified, and user authentication and security are considered. Moreover to make sure that effective communication with another layer can operate properly, we provided the services to ensure that the application can contact the database completely and process the data for the user by using internet HTTP protocols.

Table 3: Sensor Data are shown on application programming interface

Number	Raspberry Pi				Arduino			
	LDR	Humidity	Temperature	PIR	LDR	Humidity	Temperature	PIR
1	184.50	46.00	25.00	0.00	143.00	46.00	26.00	0.00
2	172.20	46.00	25.00	0.00	153.00	46.00	26.00	0.00
3	180.40	46.00	25.00	0.00	146.00	46.00	26.00	0.00
4	176.30	46.00	25.00	0.00	152.00	46.00	26.00	0.00
5	180.40	46.00	50.00	0.00	153.00	46.00	26.00	0.00
6	176.30	46.00	50.00	0.00	151.00	46.00	26.00	0.00
7	180.40	46.00	25.00	0.00	146.00	46.00	26.00	0.00
8	172.20	44.00	50.00	0.00	155.00	46.00	26.00	0.00
9	139.40	46.00	25.00	0.00	143.00	46.00	26.00	0.00
10	180.40	46.00	25.00	0.00	146.00	46.00	26.00	0.00

Table 4: Monitoring Application Program Interface Single data as the latest data

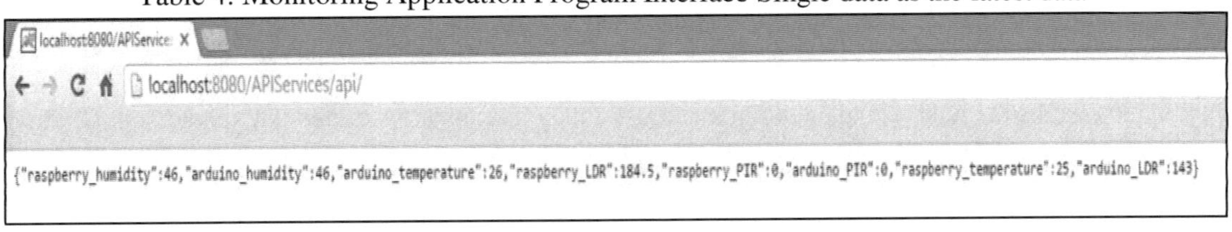

5. Conclusion

The area of the context-aware system and integration approach are provided in this research. The research touches many subjected such as sensing, control designing, integration layers, application development, et cetera. The main focus lied on applying middleware for being the bridge between physical and application layers, it has been developed to process data collected from provided sensors through prototype platforms that can be solved the problem of dependent on hardware and application. We applied and implemented integration system between the control system and context-aware computing to make them work together properly and respond to the user in real time from lower layer to application layer. An architecture of the system efficiently supports and fulfills the requirement of the system, represent and integrate contextual information, and act as the intermediary where different layers and complexities can be recovered by the proposed system. All layers are combined seamlessly and performs as completed system when they send data from raw sensor through proposed system with ease when user request data from system.

Acknowledgment

I would like to express my deeply acknowledge to those who assisted, inspired and supported me during the completion of this research such as my supervisors, colleagues. Financial support from the AUN/SEED-Net JICA project in this research and scholarship is greatly appreciated.

References

[1] Ö. Yürür, C. H. Liu, Z. Sheng, V. C. M. Leung, W. Moreno, and K. K. Leung, "Context-Awareness for Mobile Sensing: A Survey and Future Directions," *IEEE Commun. Surv. Tutor.*, vol. 18, no. 1, pp. 68–93, Firstquarter 2016.

[2] B. Schilit, N. Adams, and R. Want, "Context-Aware Computing Applications," in *Mobile Computing Systems and Applications, 1994. WMCSA 1994. First Workshop on*, 1994, pp. 85–90.

[3] J. Hong, E. Suh, and S.-J. Kim, "Context-aware systems: A literature review and classification," *Expert Syst. Appl.*, vol. 36, pp. 8509–8522, 2009.

[4] S. Lee, S. Park, and S. Lee, "A Study on Issues in Context-Aware Systems Based on a Survey and Service Scenarios," in *2009 10th ACIS International Conference on Software Engineering, Artificial Intelligences, Networking and Parallel/Distributed Computing*, 2009, pp. 8–13.

[5] G. Chen, D. Kotz, and others, "A survey of context-aware mobile computing research."

[6] W. Liu, X. Li, and D. Huang, "A survey on context awareness," in *Computer Science and Service System (CSSS), 2011 International Conference on*, 2011, pp. 144–147.

[7] K. B. Balavalad, S. S. Manvi, and A. V. Sutagundar, "Context Aware Computing in Wireless Sensor Networks," in *2009 International Conference on Advances in Recent Technologies in Communication and Computing*, 2009, pp. 514–516.

[8] T. Mo, W. Li, W. Chu, and Z. Wu, "CABS3: Context-Awareness based smart service system," in *Wireless Communications Networking and Mobile Computing (WiCOM), 2010 6th International Conference on*, 2010, pp. 1–4.

[9] A. Wood *et al.*, "Context-aware wireless sensor networks for assisted living and residential monitoring," *Netw. IEEE*, vol. 22, no. 4, pp. 26–33, Jul. 2008.

[10] F. Gui, N. Zong, and M. Adjouadi, "Artificial Intelligence Approach of Context-Awareness Architecture for Mobile Computing," in *Intelligent Systems Design and Applications, 2006. ISDA'06. Sixth International Conference on*, 2006, vol. 2, pp. 527–533.

[11] E. Kim and J. Choi, "A Context Management System for Supporting Context-Aware Applications," in *Embedded and Ubiquitous Computing, 2008. EUC '08. IEEE/IFIP International Conference on*, 2008, vol. 2, pp. 577–582.

[12] J.-W. Chang and H.-J. Lee, "Context-aware architecture for intelligent application services in ubiquitous computing," in *Semantic Computing, 2007. ICSC 2007. International Conference on*, 2007, pp. 275–281.

[13] K.-E. Ko and K.-B. Sim, "Development of context aware system based on Bayesian network driven context reasoning method and ontology context modeling," in *2008 International Conference on Control, Automation and Systems*, 2008, pp. 2309–2313.

IOP Conf. Series: Journal of Physics: Conf. Series **1195** (2019) 012021 doi:10.1088/1742-6596/1195/1/012021

[14] H. Mcheick, "Ubiquitous Computing and Context-aware Applications: Survey and Contributions," in *2016 IEEE First International Conference on Connected Health: Applications, Systems and Engineering Technologies (CHASE)*, 2016, pp. 394–397.
[15] Z. A. Hua, "Study of ubiquitous learning environment based on Ubiquitous computing," in *2010 3rd IEEE International Conference on Ubi-Media Computing*, 2010, pp. 136–138.
[16] H. A. Duran-Limon *et al.*, "Context-aware middleware for pervasive and ad hoc environments," *Context Tech Rep*, 2003.
[17] Y. Raja Vara Prasad and P. Rajalakshmi, "Context aware building energy management system with heterogeneous wireless network architecture," in *Wireless and Mobile Networking Conference (WMNC), 2013 6th Joint IFIP*, 2013, pp. 1–8.

Classification of titanium microstructure with fully convolutional neural networks

S Mongkhonthanaphon and Y Limpiyakorn

Department of Computer Engineering, Chulalongkorn University, 254 Phayathai Road, Pathumwan, Bangkok 10330, Thailand
limpyac@gmail.com

Abstract. Titanium and its alloy exhibit excellent properties for biomedical applications, especially in implant surgery. Classification of Titanium microstructure is the process in material inspection that reveals background of the material. Generally, microstructure classification is manually performed. Due to the complexity of microstructure features, expertise is required for process operation. The traditional classification by humans is time consuming and possibly error prone if the inspection is not performed by titanium microstructure experts. Deep learning is considered the revolution of computer vision to enable computers to see and perceive like humans. The technique is widely used for automatically classifying images with high accuracy. In order to reduce human inspection time during quality control, this research presents the use of a type of deep learning, Fully Convolutional Neural Networks, for pixel-wise classification in the titanium microstructure images. The dataset contains private images of titanium samples taken by SEM microscopes. As the available training dataset is small, data augmentation using elastic deformations is applied for increasing the accuracy of the model. Constructed with the U-net architecture, the network achieves good performance with the pixel accuracy of 92.67% and mean IoU of 71.30%.

1. Introduction

Titanium and its alloys have become very important commercially over the past fifty years due to their low density, good mechanical properties, good strength-to-weight ratio, good corrosion resistance and good biocompatibility. Surgery, especially in implant surgery uses titanium alloys as implant material. The manufacturing process for implant material therefore has to be able to build complex shape. The challenge is how to make this titanium as precise as possible.

Since 2006, selective laser melting (SLM) has been introduced with the ability to make complex shape, reduce fabrication time and reduce material waste from the producing process. Unfortunately, as-built state titanium parts from SLM are unable to achieve the great performance. Their toughness and ductility are not high enough to be used as implant material because they cannot take too much load. To improve the toughness and ductility, heat treatment is needed. To improve some mechanical behaviour, Harr and Becker [1] experimented the heat treatment process on Ti-6Al-4V (6% aluminium, 4% vanadium titanium alloy), produced by SLM method. The result showed a significant increase of %elongation for the samples that have been heated above 750°C (dissolution temperature). An as-built state of Ti-6Al-4V consists only the Alpha phase. The dissolution temperature is the temperature that the Alpha phase is about to start dissolve into Beta phase. The other matters are time and method of cooling that can affect the microstructure morphology. Mostly of microstructure visualization are manually performed by metallurgists including phase classification. In this work, deep learning is

proposed as a means to automate the classification of Ti-6Al-4V microstructure in order to reduce the time spent on inspection for quality control.

Convolutional Neural Networks (CNNs) have been one of the most successful for image classification. In 1989, the first CNN was proposed by LeCun et al. [2] who used the network to recognize patterns on hand-written digits dataset. In 2012, Krizhevsky et al. [3] adapted LeNet into the breakthrough AlexNet with 7-stacked layers that won the ImageNet Large Scale Visual Recognition Challenge (ILSVRC) in that year. Moreover, AlexNet drastically improved the accuracy of the competition by 10% from 2011. This phenomenon had drawn interest into the field of Deep Learning.

Semantic segmentation is one of the key problems in the field of computer vision, and it can be applied to a variety of applications including self-driving car, medical diagnosis, virtual reality etc. The well-known method in semantic segmentation, so called Fully Convolutional Neural Networks (FCNN), were proposed by Long et al. in 2015 [4]. Compared to the original CNNs, FCNNs are the model containing upsampling layer instead of fully connected layer. Therefore, FCNNs can deliver the output as the same resolution as the input image. Another difference is Skip Connections that skip the information of the shallower layers to the deeper layers. This method improves further accuracy by fusing the information before the final prediction layer.

2. Related work

Thus, it is such a challenge to make computer vision recognize the patterns in microstructure. In the field of metallurgy, there are several instruments for imaging, such as optical microscope (OM) and scanning electron microscope (SEM). The SEM method yields higher resolution in the image compared to the OM method, but the operating cost is more expensive. Gola et al. [5] proposed steel microstructure classification by machine learning methods, using Support Vector Machine (SVM) as classifier to predict on steel microstructure dataset. The dataset consists of steel images that taken by OM method. All images are categorized into three classes: pearlite, martensite and bainite. The result has shown the average test accuracy around 80% on the author's test dataset. Azimi et al. [6] applied the method of Gola et al. to their dataset, which was categorized into four classes: pearlite, tempered martensite, martensite, and bainite. However, the result showed only the accuracy of 48.89%. Azimi et al. thus proposed steel microstructure classification by deep learning methods, using FCNN [4] to classify every pixel in whole image and predict the class by max-voting scheme. The model achieved 93.94% classification accuracy and 67.84% mean IoU. The publication of Iglovikov and Shvets [7] in 2018 presented the technique of fine-tuning to increase the accuracy with the pretrained network previously trained on a large dataset. The pretrained network was required to freeze the weights for almost the whole of convolution bases except for the very last one or two layers. Therefore, when training the new classifier, the new data will run through the network to adjust the weights of last convolution base layer. The pretrained network could work on contracting part of the U-net architecture, considered as original CNN.

3. Method

3.1. Architecture

The U-net architecture is built from FCNN concept to perform image segmentation, and it successfully works on biomedical image segmentation as the winner of ISBI cell tracking challenge 2015 [8]. The architecture is named after U alphabet as it consists of two parts concatenating like U-shape as illustrated in figure 1. The architecture can be seen as two divided parts, the part on left-side is contracting path where the architecture encodes the input image into feature map by using convolution and max-pooling operations. The right-side part is expansive path where the architecture decodes the feature map into output segmentation map by using upsampling. The addition idea of this architecture is that the expansive path also has huge feature channels, that is different from the original FCNN [4]. As when the upsampling is used, the feature map channels are halved so that the channels in the deeper layers tend to be smaller. But U-net uses concatenation method to crop and copy the feature maps in contracting

IOP Conf. Series: Journal of Physics: Conf. Series **1195** (2019) 012022 doi:10.1088/1742-6596/1195/1/012022

path to merge with the feature maps in the expansive path. In summary, the constructed model has 23 convolutional layers (eighteen 3x3 convolutions, four 2x2 upsampling convolutions, one 1x1 convolution) and 4 max-pooling operations.

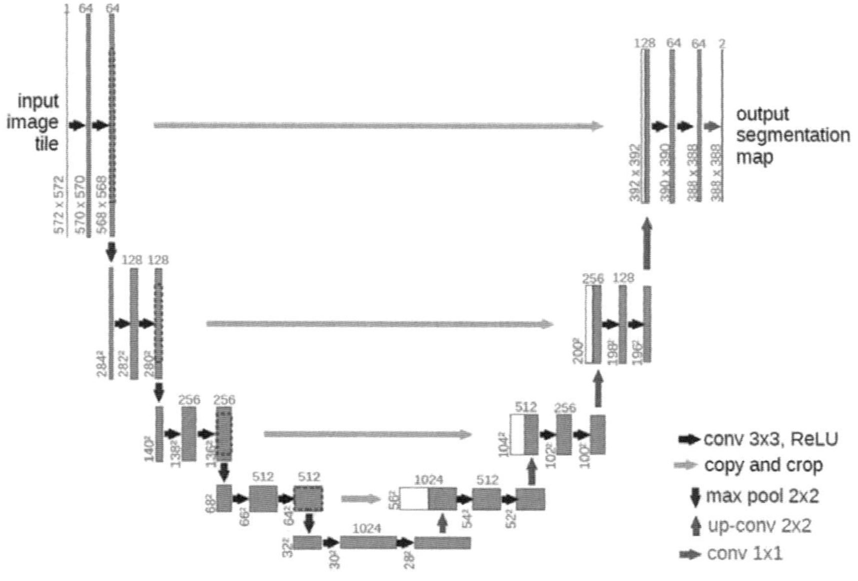

Figure 1. U-net architecture [8].

3.2. Dataset

The dataset contains images of titanium samples taken by SEM microscopes. These titanium samples are produced by SLM method and heated to transform the as-built microstructure into dual-phase microstructure. (Alpha-Beta) The dataset contains 96 images with 512x512 pixels. Each image has a corresponding ground-truth image that supervises the classes to the model. The ground-truth images are denoted for Alpha phase (black) and Beta phase (white) in the images. The dataset is split into 66 training images and 30 test images. As the available dataset is small, data augmentation using elastic deformations is applied for increasing data to the size of 264 images. Figure 2 illustrates an example of the augmented training image (on the right) that is the output from pixel shifting by shifting 5% of total width to the left and shifting 5% of total height lower. Associated with each augmented image, the ground truth image is also created using the same pixel shifting. Figure 3 illustrates an example of the augmented ground-truth image (on the right) resulting from pixel shifting. This step can be carried out with Keras [9] using ImageDataGenerator instance.

3.3. Training

The network produces an output of a 3D matrix with the number of channels equivalent to the number of classes. Since there are only two classes (Alpha and Beta), ground truth images are thus preprocessed into binary color images by thresholding from grayscale images. Thresholding is manually performed by replacing each pixel in an image with a black pixel if the image intensity is less than some fixed constant, manually chosen by visualizing the histogram color map. The images of titanium alloys taken by SEM microscope are used as input and all binary ground-truth images are used to supervise the network in the output section. For training, Adam optimizer [10] is used with 10^{-4} learning rate and no weight decay. The loss function is computed by using binary-cross entropy function. To predict the output of the last layer for every pixel, the sigmoid function is applied to score the posterior probability then choosing the class for that pixel with the highest posterior score.

IOP Conf. Series: Journal of Physics: Conf. Series **1195** (2019) 012022 doi:10.1088/1742-6596/1195/1/012022

3.4. Implementation

All of the implementations have been done with Tensorflow and Keras [9] on a single GPU Nvidia GTX1060 (6gb). The network was trained with 15 epochs. Training time is about 5 hours from scratch.

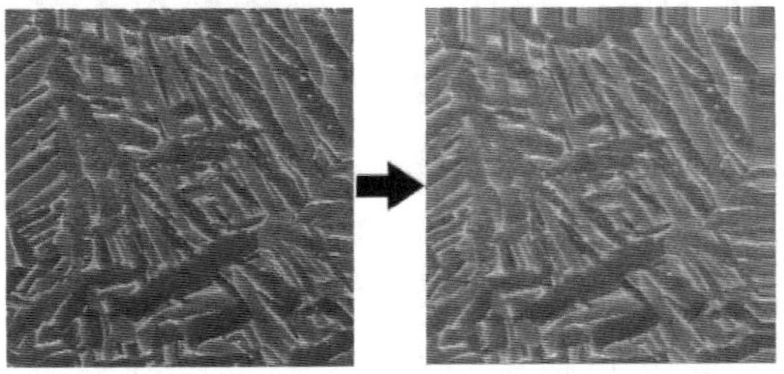

Figure 2. Example of augmented training image with pixel shifting.

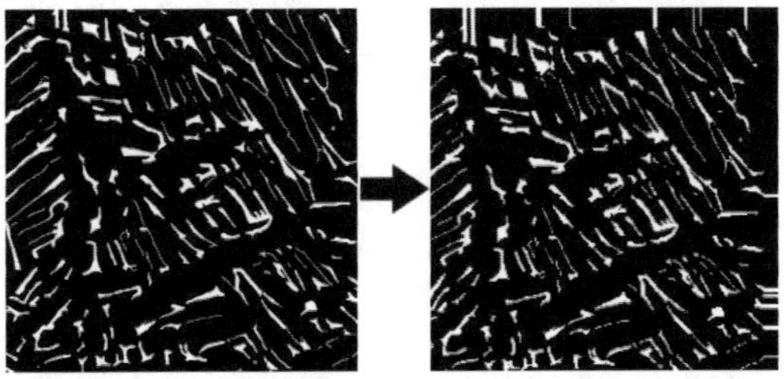

Figure 3. Example of augmented ground truth image with pixel shifting.

4. Result

Once the training has been completed, the training weights are then saved in .hdf5 format. The size of file containing the values of training weight is around 375 MB. To predict the class in the test dataset it is required to load the model architecture and the weight file. The example result is shown in figure 4 where (a) is the original image of Ti-6Al-4V microstructure, (b) is the associated ground truth image, (c) is the predicted image on augmented dataset, and (d) is the predicted image on original dataset.

The evaluation of the performance of semantic segmentation includes four widely used metrics: 1) pixel accuracy, 2) mean accuracy, 3) mean of intersection over union (IoU), and 4) frequency weighted of IoU. Given that n_{ij} denotes the number of pixels of class i-th predicted to belong to class j; n_{cl} is the number of different classes included in ground truth segmentation; and t_i is the total number of pixels of class i-th in ground truth segmentation. The calculation of the four metrics is defined as in equation (1), (2), (3), and (4), respectively.

$$\text{pixel accuracy} = \frac{\sum_i n_{ii}}{\sum_i t_i} \quad (1)$$

$$\text{mean accuracy} = \frac{1}{n_{cl}} \frac{\sum_i n_{ii}}{t_i} \quad (2)$$

$$\text{mean IoU} = \frac{1}{n_{cl}} \sum_i \frac{n_{ii}}{(t_i + \sum_j n_{ji} + n_{ii})} \quad (3)$$

$$\text{frequency weighted IoU} = \sum_k \frac{1}{t_k} \sum_i \frac{t_i n_{ii}}{(t_i + \sum_j nji - n_{ii})} \quad (4)$$

The summary of model performance is detailed in table 1reporting all the measures on two datasets: with or without data augmentation. In case of using data augmentation and Adam optimizer strategy, the network achieves the best pixel accuracy of 92.67%, mean accuracy of 80.39%, mean IoU of 71.30% and frequency weighted IoU of 87.24%, computed by comparing the prediction images with the ground-truth images prepared for testing. The results on non-data augmentation achieve lower performance, that is, the pixel accuracy of 83.09%, mean accuracy of 67.27%, Mean IoU of 65.12% and frequency weighted IoU of 82.03%. Observing that data augmentation significantly increases the accuracy of the network.

Table 1. The effect of data augmentation on FCNN.

Training strategy	pixel accuracy	mean accuracy	mean IoU	frequency weighted IoU
no data augmentation	0.8309	0.6727	0.6512	0.8203
data augmentation	**0.9267**	**0.8039**	**0.7130**	**0.8724**

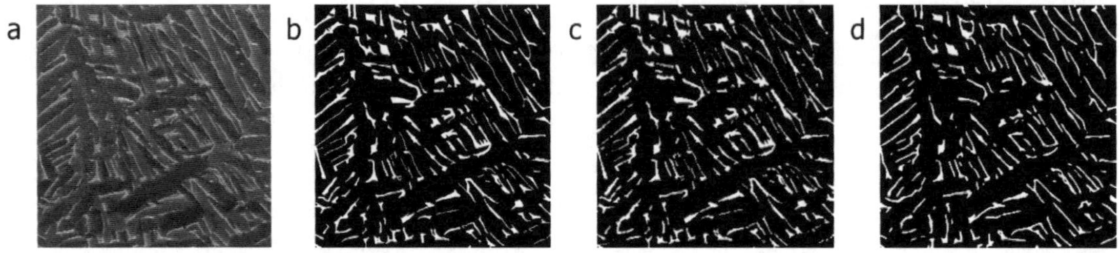

Figure 4. (a) example of test image, (b) ground truth image of for test, (c) predicted image using augmented dataset, (d) predicted image using non-augmented dataset.

5. Conclusion

The unique properties of Titanium alloys have a potential to solve problems encountered by current biomedical materials. Improvements in both the quality and quantity of the implant material have made this treatment modality very promising and highly practiced in today's era. Since manual inspection of Titanium microstructure is resource consumption, this paper presents an approach to applying deep learning for the classification of the Ti-6Al-4V microstructure. The FCNN classifier is built on U-net architecture to classify the phases (Alpha or Beta) in titanium microstructure images. As the small dataset is available, data augmentation with elastic deformations is performed for increasing data to prevent model overfitting. The performance of the network achieves the pixel accuracy of 92.67% and mean IoU of 71.30%. Further investigation of other techniques would be carried out to improve the network performance. For example, integrating Fine-tuning into the U-net architecture. The technique requires the pretrained network previously trained on a large dataset.

References

[1] Terhaar, G., M., Becker, T., H. (2018) Selective Laser Melting Produced Ti-6Al-4V: Post-Process Heat Treatments to Achieve Superior Tensile Properties. Journal of Physics, Volume 371, conference 1

[2] LeCun, Y., Boser, B., Denker, J., S., Henderson, D., Howard, R., E., Hubbard, W., Jackel, L., D. (1989) Backpropagation applied to handwritten zip code recognition. Neural Computation 1(4), 541-551

[3] Krizhevsky, A., Sutskever, I., Hinton, G., E., (2012) *Imagenet classification with deep convolutional neural networks*. In NIPS, 1106-1114

[4] Long, J., Shelhamer, E., & Darrell, T. (2015). Fully convolutional networks for semantic segmentation. IEEE Conference on Computer Vision and Pattern Recognition (CVPR)

[5] Gola, J., Britz, D.,Staudt, T., Winter, M., Schneider, S., A., Ludovici, M., Mucklich, F. (2016) Advanced microstructure classification by data mining methods. International Conference on ThermoMechanical Processing (TMP).

[6] Azimi, S., Britz, D., Engstler, M., Fritz, M. and Mücklich, F. (2018) Advanced Steel Microstructural Classification by Deep Learning Methods. Nature-Scientific Reports

[7] Iglovikov, V., Shvets, A. (2018) *TernausNet: U-Net with VGG11 Encoder Pre-Trained on ImageNet for Image Segmentation*. IEEE Conference on Computer Vision and Pattern Recognition (CVPR)

[8] Ronneberger, O., Fischer, P., Brox, T. (2015) U-Net: Convolutional Networks for Biomedical Image Segmentation, MICCAI 2015

[9] Chollet, F. (2017) Deep learning with Python. New York: Manning

[10] Kingma, P., D., Ba, L., J. (2015) *Adam: A method for stochastic optimization.* In ICLR 2015

Numerical/multiphysical investigation of shrinking hole at notch tip under high electric energy

Thomas Jin-Chee Liu

Department of Mechanical Engineering, Ming Chi University of Technology, Taishan, New Taipei City, Taiwan

E-mail: jinchee@mail.mcut.edu.tw

Abstract. In this paper, the shrinking hole creation at the notch tip is simulated using the finite element method (FEM) and element birth and death (EBD) technique. The removed material, stress, temperature, and electric current density are obtained for estimating the effects of the hole creation. Also, the variation of the shrinking hole size is discussed by the numerical/multiphysical simulation. When the melting material is removed, it shows rational stress values. The stress field is affected by the shrinking hole size and it must be considered in the analysis.

1. Introduction

The Joule heating effect is an electro-thermal coupling phenomenon. According to past studies [1-6], under the electric load, the electric concentration, Joule heating and local hot spot exist at the crack/notch tip in the conductive material. In Fig. 1, the electric current density concentrates at the notch tip and then a hot spot occurs.

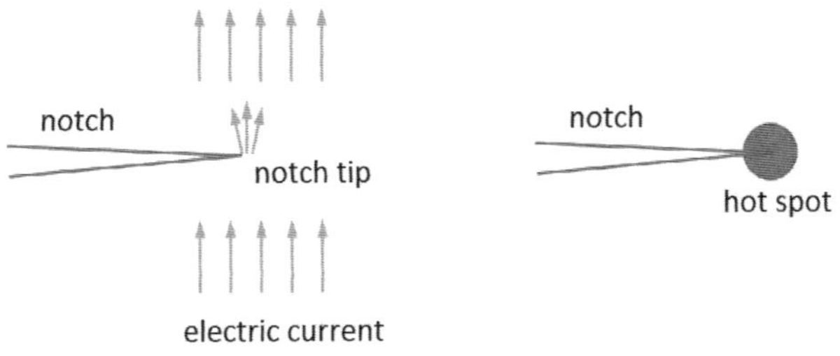

Figure 1. Joule heating effect at notch tip.

Under high electric energy, the local hot spot at the notch tip can melt and then a shrinking hole may produce. In the author's past studies [7,8], numerical and experimental results present the hot spot, melting area, solidification, shrinking hole and heat affected zone around the notch tip under high electric current. These results provide the concepts for increasing the fatigue life or stopping the crack propagation [7,8]. However, it was found that the production of the shrinking hole is a complicated case. As shown in Fig. 2, the author's experimental study shows different hole size under the same loading and boundary conditions [8].

In this paper, the shrinking hole creation at the notch tip will be simulated using the finite element method (FEM) and element birth and death (EBD) technique of the ANSYS software. The removed material, stress, temperature, and electric current density will be obtained for estimating the effects of the hole creation. Especially, the variation of the shrinking hole size such like Fig. 2 will be discussed by FEM and EBD.

Figure 2. Variation of shrinking hole size at notch tip (applied electric current = 600 A, operating time = 0.3333 s) [8].

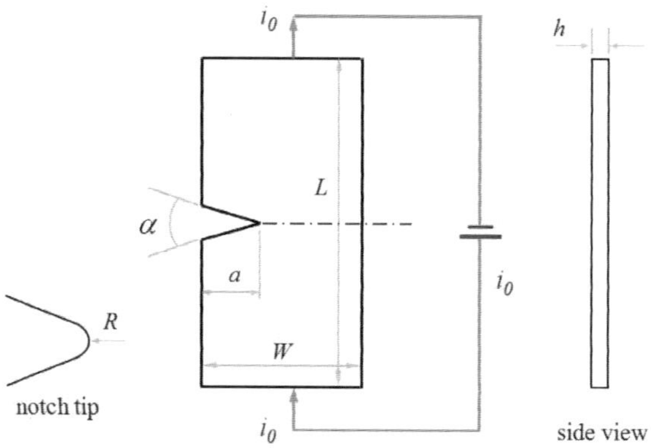

Figure 3. Case study.

2. Case study

For this case study, the sample is the same as refs. [7,8]. As shown in Fig. 3, the stainless steel strip has the dimensions $W \times L \times h$. The notch length and notch angle are respectively denoted as a and α. The round notch tip radius is R_t. The steel strip is subjected to a direct current (DC) i_0. The material type of the strip is SUS 304 stainless steel and the material constants are listed in Table 1 [9,10]. The elasto-plastic stress-strain relation with the tangent modulus E_T=0.4E is considered in the FEM.

The geometric properties are L=0.1 m, W=0.008 m, a=0.004 m, h=0.001 m, α=20°, and R_t=0.0002 m. The electric load i_0 is 600 A. The following conditions are considered:

(1) Three-dimensional analysis
(2) Electro-thermo-structural coupled-field analysis

(3) Transient heat transfer (initial condition: 25 °C)
(4) Convection on air/steel interface (natural convection coefficient: 25 W/(m^2-K), ambient temperature: 25 °C)
(5) Steady DC/RMS electric current

Table 1. Material data of SUS 304 stainless steel [9,10].

Temperature (°C)	Young's modulus E (GPa)	Yielding strength S_Y (MPa)	Coefficient of thermal expansion α (1/°C)	Thermal conductivity k (W/ m-°C)	Specific heat C_p (J/ kg-°C)	Resistivity ρ (Ω-m)
20	-	-	16.3×10^{-6}	16.2	456	7.2×10^{-7}
27	193	290	-	-	-	-
90	-	-	-	-	490	7.8×10^{-7}
93	192	-	-	-	-	-
100	-	-	16.3×10^{-6}	16.2	-	-
149	187	182	-	-	-	-
200	-	-	16.7×10^{-6}	-	532	8.6×10^{-7}
204	183	-	-	-	-	-
260	179	150	-	-	-	-
300	-	-	17.1×10^{-6}	-	-	-
316	177	-	-	-	-	-
320				-	557	9.5×10^{-7}
371	170	134	-	-	-	-
400	-	-	17.6×10^{-6}	-	-	-
427	166	-	-	-	-	-
430	-	-	-	-	574	10.2×10^{-7}
482	160	125	-	-	-	-
500	-	-	18.0×10^{-6}	21.5	-	-
538	155	-	-	-	-	-
540	-	-	-	-	586	10.8×10^{-7}
593	150	113	-	-	-	-
600	-	-	18.3×10^{-6}	-	-	-
649	145	-	-	-	-	-
650	-	-	-	-	599	11.4×10^{-7}
700	-	-	19.0×10^{-6}	-	-	-
704	141	95	-	-	-	-
760	134	-	-	-	620	11.8×10^{-7}
800	-	-	20.0×10^{-6}	-	-	-
816	125	68	-	-	-	-
870	-	-	-	-	645	12.5×10^{-7}

Poisson's ratio ν= 0.29, density β= 8000 kg/m^3, melting point = 1400 °C.

3. Numerical methods

The finite element equations of the thermo-electro-structural coupled-field analysis are listed as follows [11]:

$$\begin{bmatrix} \mathbf{M} & 0 & 0 \\ 0 & 0 & 0 \\ 0 & 0 & 0 \end{bmatrix}\begin{bmatrix} \ddot{\mathbf{U}} \\ \ddot{\mathbf{T}} \\ \ddot{\mathbf{V}} \end{bmatrix} + \begin{bmatrix} \mathbf{C} & 0 & 0 \\ \mathbf{C}^{\mathrm{tu}} & \mathbf{C}^{\mathrm{t}} & 0 \\ 0 & 0 & 0 \end{bmatrix}\begin{bmatrix} \dot{\mathbf{U}} \\ \dot{\mathbf{T}} \\ \dot{\mathbf{V}} \end{bmatrix} + \begin{bmatrix} \mathbf{K} & \mathbf{K}^{\mathrm{ut}} & 0 \\ 0 & \mathbf{K}^{\mathrm{t}} & 0 \\ 0 & 0 & \mathbf{K}^{\mathrm{v}} \end{bmatrix}\begin{bmatrix} \mathbf{U} \\ \mathbf{T} \\ \mathbf{V} \end{bmatrix} = \begin{bmatrix} \mathbf{F} \\ \mathbf{Q} \\ \mathbf{I} \end{bmatrix} \tag{1}$$

where **U**, **T**, **V**, **F**, **Q** and **I** are the vector forms of the displacement, temperature, electric potential, force, heat flow rate and electric current, respectively. The material constant matrices **M**, **C**, $\mathbf{C}^{\mathrm{t}}$, $\mathbf{C}^{\mathrm{tu}}$, **K**, $\mathbf{K}^{\mathrm{t}}$, $\mathbf{K}^{\mathrm{ut}}$ and $\mathbf{K}^{\mathrm{v}}$ are the structural mass, structural damping, thermal specific heat, thermo-structural damping, structural stiffness, thermal conductivity, thermo-structural stiffness and electric conductivity, respectively. The coupled heat flow matrix **Q** contains the effects of the thermal loading

and electrical Joule heating. $\mathbf{C}^{tu}$ and $\mathbf{K}^{ut}$ are thermo-structural coupled terms. Equation (1) is a directly coupled nonlinear equation which is solved using the Newton-Raphson iterative method.
As shown in Fig. 4, the finite element model is established using ANSYS software. The solid element SOLID226 is adopted to construct the numerical mesh. The nodal degrees of freedom of SOLID226 are the displacements, temperature and electric potential.

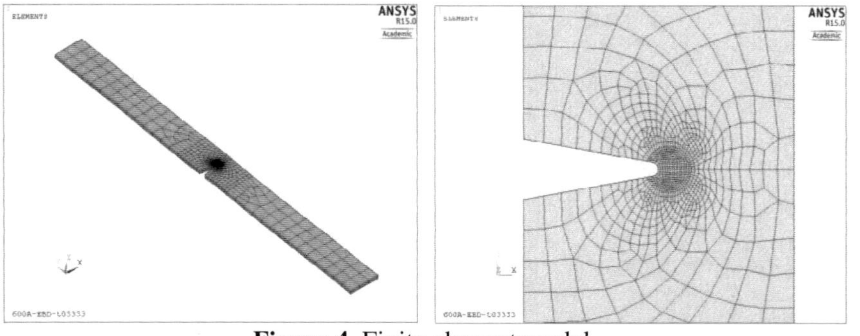

Figure 4. Finite element model.

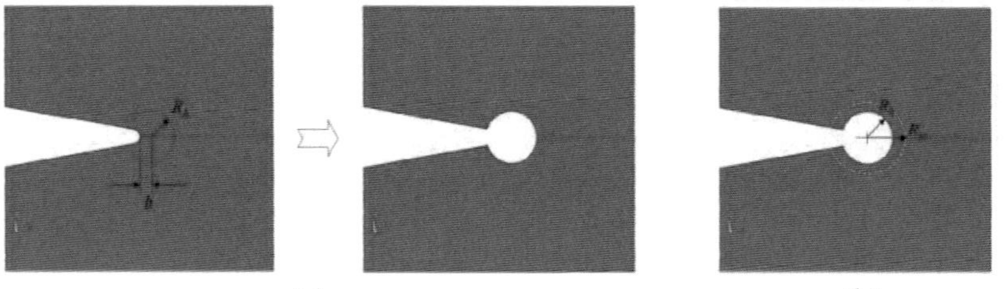

(a) (b)

Figure 5. Element birth and death (EBD) technique. (a) Concept, (b) Assumption of R_h.

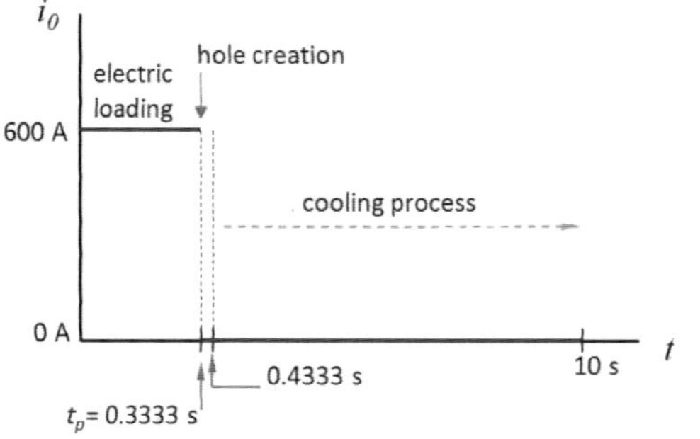

Figure 6. Loading history.

The element birth and death (EBD) technique of the ANSYS [11] is adopted to simulate the shrinking hole creation. The concept of EBD is plotted in Fig. 5. The circular area (radius=R_h) of the material is removed to simulate the shrinking hole creation. The time of hole creation is at the end time of the electric load. After that, the electric load is removed and the structure is subjected to the cooling condition. Figure 6 shows the loading history of this study. The applied electric current i_0= 600 A. The operating time is t_p= 0.3333 s.

Referring to the photos in Fig. 2, it is difficult to define the accurate shrinking hole radius R_h. As shown in Fig. 5(b), the relation $R_h=KR_m$ is assumed, where K and R_m are the scale factor and the radius of the melting circle, respectively. R_m must be defined from the finite element results.

4. Numerical results

4.1 Typical results

Based on the loading history in Fig. 6, the numerical results under K=1 are shown in this section. The condition K=1 expresses that the melting area (T >1400 °C) is the same as the shrinking hole size.

In Figs. 7 and 8, the history of the electric current density and temperature filed are plotted around the notch tip. After the time t=0.3333 s, the circular region is removed to simulate the production of the shrinking hole. The heat source at the notch tip is induced by the electric concentration and Joule heating.

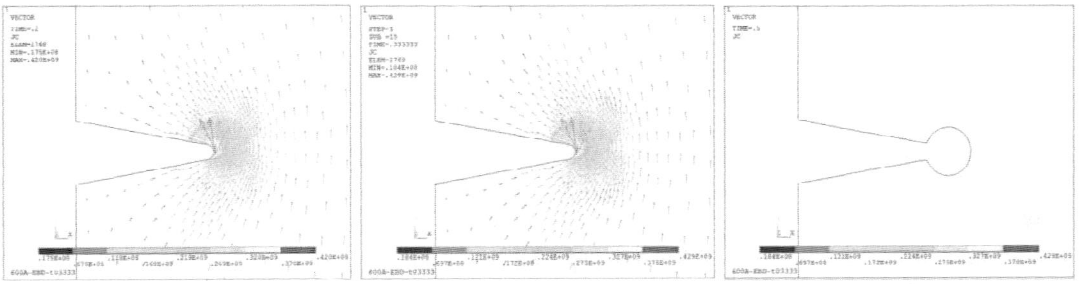

t = 0.2 s t = 0.3333 s t = 0.5 s

Figure 7. History of electric current density (units: A/m^2).

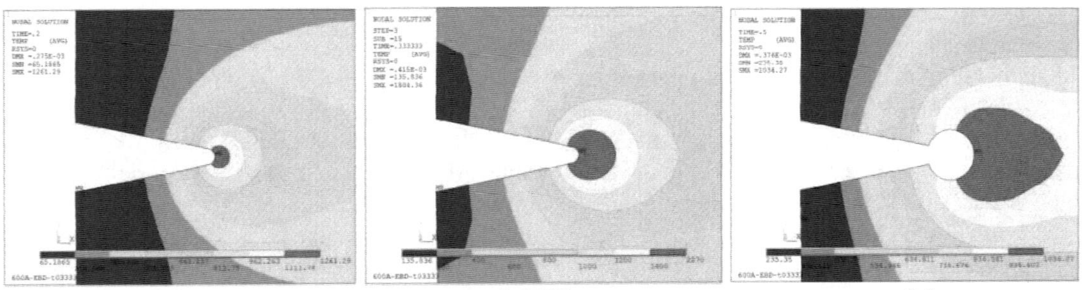

t = 0.2 s t = 0.3333 s t = 0.5 s

Figure 8. History of temperature filed (units: °C).

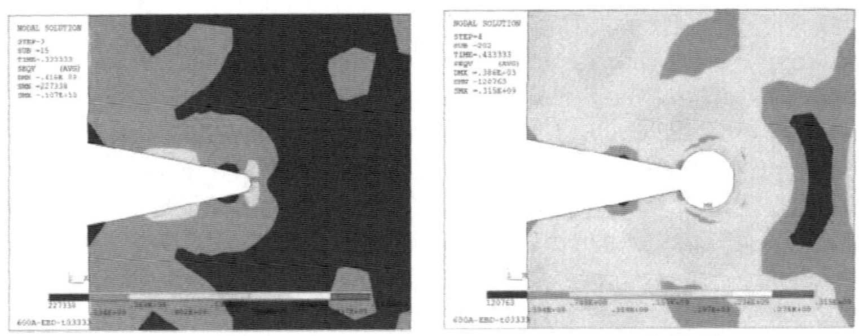

t = 0.3333 s t = 0.4333 s

Figure 9. von Mises stress (units: Pa).

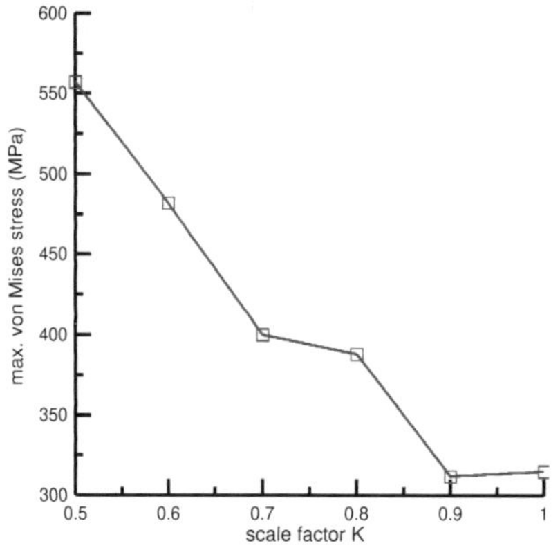

Figure 10. Hole size effects on von Mises stress

In Fig. 9, the von Mises stresses at t = 0.3333 s and t = 0.4333 s are plotted. At t = 0.3333 s, the highest stress value is not rational because the melting material is not removed. However, it shows rational results when the shrinking hole is considered at t = 0.4333 s.

4.2 Shrinking hole size

Figure 10 shows the maximum von Mises stresses around the shrinking hole with K=0.5~1 when t = 0.4333 s. The stress value of K=0.5 is higher than others. It means that the stress field is affected by the shrinking hole size and it must be considered in the analysis.

From the results in Fig. 10, the smaller K, i.e. smaller shrinking hole, induces higher stress value. This phenomenon is due to the thermal deformation and stress. The condition of smaller K means that more material is kept around the shrinking hole and it makes higher thermal stress field.

5. Conclusions

The real case of the melting/cooling process is very complicated so that the numerical simulation can not consider complete situations of the shrinking hole. However, the EBD method provides a better solution to analyze the shrinking hole at the notch tip. When the melting material is removed, it shows rational stress values. The stress field is affected by the shrinking hole size and it must be considered in the analysis. The smaller shrinking hole induces higher stress value.

References

[1] Parton V Z and Kudryavtsev B A 1988 *Electromagnetoelasticity* (New York: Gordon and Breach)
[2] Fu Y M, Bai X Z, Qiao G Y, Hu Y D and Luan J Y 2001 *Mater. Sci. Tech.* **17** 1653
[3] Qin Z, Librescu L and Hasanyan D 2007 *J. Therm. Stress.* **30** 623
[4] Cai G X and Yuan F G 1999 *Int. J. Fract.* **96** 279
[5] Liu T J C 2011 *Eng. Fract. Mech.* **78** 666
[6] Liu T J C 2014 *Eng. Fract. Mech.* **123** 2
[7] Liu T J C, Cheng C D, Tseng J F, Chen L W and Chen P H 2016 *Procedia CIRP* **42** 679
[8] Liu T J C, Tseng J F and Cheng C D 2018 *MATEC Web of Conferences* **151** 01004
[9] MatWeb LLC 2018 *Online Materials Information Resource – MatWeb* (http://matweb.com/)
[10] American Iron and Steel Institute 2006 *High-Temperature Characteristics of Stainless Steels*
[11] ANSYS Inc 2011 *ANSYS 14.0 Mechanical APDL Theory Reference*

Acknowledgements

The author would like to thank the Ministry of Science and Technology in Taiwan for the financial support under contract numbers MOST 104-2221-E-131-030 and MOST 105-2221-E-131-014.

IOP Conf. Series: Journal of Physics: Conf. Series **1195** (2019) 012024 doi:10.1088/1742-6596/1195/1/012024

CAD based design of a high energy absorption metal oxide varistor

Razvan Petrenci[1], Mihaela Frigura-Iliasa[1], Flaviu M. Frigura-Iliasa[2], Marius Mirica[2], Lia Dolga[3], Hannelore E. Filipescu[3]

[1]Politehnica University/Power Systems Department, Timisoara, Romania
[2]National Institute for Res. and Development in Electrochemistry and Condensed Matter/LERF, Timisoara, Romania
[3]Politehnica University/Mechatronics Department, Timisoara, Romania

razvanclaudiupetrenci@yahoo.com; mihaela.frigura@gmail.com; flaviu.frigura@upt.ro; miricamc@yahoo.com; lia.dolga@upt.ro; hannelore.filipescu@upt.ro

Abstract. Metal Oxide Varistors are a very common power electronic device, applied for efficient overvoltage protection at any voltage level. This piece of equipment has a high non-linear current response function of the applied voltage, and, it provides a relatively high heat absorption capacity in case of accidental overvoltage shocks. The crossing response current is clearly activated by temperature of that device, and, by consequent, overheating could be disastrous. Actual researches must be carried out both for a new more performant material as well as for new technical solutions for the design of all equipment integrating them, by studying heat extraction and heat transfer inside a new complex varistor device. Our article proposes a totally new device, used basically for low voltage applications, having a supplementary metal mass added to the body of that varistor, shaped as small disk. It actions like a heat pump immediately after the voltage shock and as additional radiators at the end of the heating process caused by a transitory overvoltage. A CAD solution combined with a finite element model, followed by some experimental results are also presented, for confirming the performance of that newly design. By placing additional metal alloy masses inside a new varistor structure it will have a higher heat pumping and dissipation capability, in order to reduce temperature stress and all aging effects.

Index Terms—Metal Oxide Varistors, CAD model, low voltage, energy absorption

1. Introduction

Nowadays, Metal Oxide Varistors are often used for overvoltage protection equipment on a large industrial scale. All low voltage sensitive electronic devices are totally exposed to a lot of overvoltage aggressions, caused mostly by lightning (as transitory overvoltage) or caused by internal power network faults (as permanent overvoltage).

The Metal Oxide Varistors are made of a mixture of semiconductor ceramic materials, most of them oxides.

IOP Conf. Series: Journal of Physics: Conf. Series **1195** (2019) 012024 doi:10.1088/1742-6596/1195/1/012024

The relation between their crossing current and their applied voltage is essentially non-linear and their response time is extremely short, faster than the dielectric breakdown caused by the residual lightning voltage stroke on the sensitive and protected equipment. They have also a relatively high level of energy absorbtion capacity, but, having a risk of overheating for long time, due to their current which is highly influenced by temperature. Controlling their temperature is crucial for maintaining them in long service [1].

The current crossing throughout a varistor is highly depending on temperature, as for all semiconductor materials. In case of consecutives lightning strokes applied to the power grid, or in case of a longtime overvoltage, we noticed also an increased risk of overheating for the varistor itself. When temperature increases (also due to environmental causes), the current increases too, due to electrical resistance reduction. An avalanche reaction could possibly appear in this case, with fatal and expensive consequences both for the protection as well as for the protected fragile piece of equipment [2].

Heat quantity produced inside the mass of that varistor is important in any case, and, by consequent, the design of an adequate technical solution, in order to control heat dissipation or evacuation, is mandatory for any varistors involved, at all voltage levels.

There are many practical and simple technical solutions applied mostly to increase the energy absorption capability (capacity) in case of a temporary lightning shock (radiators, coating, etc.), not all of them being so effective. Today there is no analytical model which could describe a complex overvoltage incident, both from the electrical as well as the thermal point of view. Only the experimental procedures are suitable for this task.

This article describes only models and measurements made in case of the permanent overvoltage service. This permanent (long term) overvoltage service is not completely described in today's literature. In this situation, the varistor is exposed to a longtime accidentally occurring overvoltage, not so high, but enough destructive for the sensitive protected equipment;

Heat (heat power, per time unit) produced inside the varistor (only by Joule effect) is given by:

$$P_{dez} = U \cdot I = U \cdot A(U) \cdot T^2 \cdot e^{\frac{q_e \cdot \Phi(U)}{k \cdot T}} \tag{1}$$

Heat (heat power) evacuated by convection and radiation to the environment is given by [3-5]:

$$P_{dis} = \alpha \cdot S_l \cdot (\theta - \theta_a) \tag{2}$$

where:

- I – crossing current;
- U - applied voltage;
- S_l – complete external heat exchange surface;
- $\Phi(U)$ – voltage height of semiconductor's junction potential barrier;
- Θ – varistor own temperature;
- Θ_a – environmental temperature;
- T – absolute temperature, on Kelvin scale;
- k – Boltzmann's constant;
- $A(U)$- a parametric function depending on voltage and electric field intensity;
- q_e – electric charge of an electrone;
- α – the combined convective and radiative heat exchange coefficient;

There is a typical intersection of these two graphs (P_{dis} and P_{dez}) given by (1) and (2), providing two equilibrium points. The situation in which these two graphs do not intersect each other at all is the situation of a permanent fatal overheating regime (when heat produced inside the varistor is too high to be evacuated in the environment).

Thermal stability (and energy adsorption), for a specified varistor, was analyzed only by taking in consideration the temperature of that varistor as the main perturbation, as a direct consequence of the overvoltage exposure [6-8].

IOP Conf. Series: Journal of Physics: Conf. Series **1195** (2019) 012024 doi:10.1088/1742-6596/1195/1/012024

2. Additional metal mass principle

As previously explained, the most important issue concerning heat absorbtion capacity is to maintain the varistor inside the so-called thermal stability envelope, by carefully limiting its temperature.

The high voltage exposure of a certain varistor also to a high energy shock wave is caused by a short-time overvoltage, (process considered as adiabatic, due to its extremely short time). It means that the whole energy $Q=W$, produced by Joule effect effect remains stored inside its own mass, causing an important increase of its natural temperature $\Delta\theta_1$. It's necessary that this temperature increase to not over-pass the temperature stability equilibrium limit, so all heat produced inside the varistor could be dissipated in the environment.

The equation of for the varistor temperature increase $\Delta\theta_1$ due to the Q heat stored inside, is:

$$W = Q = m_v \cdot c_v \cdot \Delta\theta_1 \tag{3}$$

In practice, after a brief adiabatic process, the whole heat produced inside that varistor by Joule effect, Q, is stored and kept inside the varistor mass m_v, having the c_v specific mass heat. Heat quantity Q is given by the overvoltage specificity, so it could not be changed, and c_v is a specific material parameter, which could be slightly increased for higher heat absorption, but with drastic consequences, mostly for the electrical properties of that material. It is considered as a specific constant for a certain varistor type/ material.

An efficient and original technical solution (patented in Romania, by the authors), which is strongly recommended for thermal stability control of a certain varistor, consists in attaching some additional masses (made of a conductor material, metal or metal alloy) on the varistor itself These additional metal based masses must have an excellent thermal and electrical contact with the varistor. The main principle of this technical solution is briefly synthesized in Fig. 1.

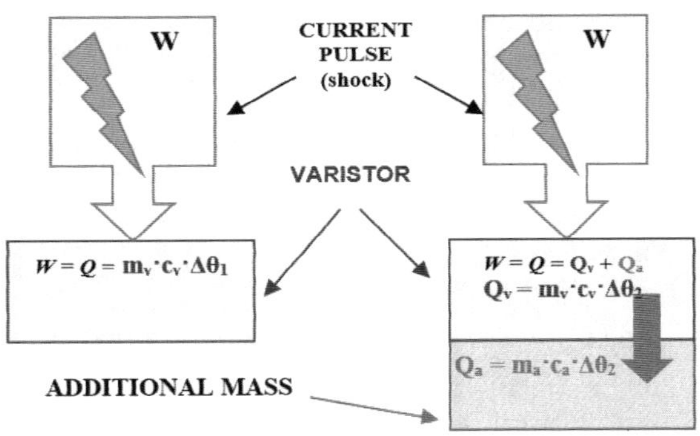

Figure 1. The additional mass solution.

The solution of this additional metal mass consists essentially in dividing the heat quantity produced by Joule effect inside the active part of the varistor, $Q=W$, in two different fractions. One of these fractions, called Q_v ($Q_v = m_v \cdot c_v \cdot \Delta\theta_2$), remains stored inside the varistor, and the other fraction, called Q_a ($Q_a = m_a \cdot c_a \cdot \Delta\theta_2$) is naturally sent to and stored inside the additional mass which is well thermo-coupled with the varistor. Of course, m_a is the additional metal mass and c_a is its own specific mass heat.

In this situation, the corresponding relation is:

$$W = Q = Q_v + Q_a = (m_v \cdot c_v + m_a \cdot c_a) \cdot \Delta\theta_2 \tag{4}$$

Equation (4) is valid and correct because:

- The electrical resistance of that additional mass is nine times smaller than the smallest varistor resistance obtained for the best conduction mode, and, by consequent, the additional mass is a passive conductor, which is not modifying any electric property of the assembly. The heat

produced by Joule effect inside this additional mass is totally negligible, and, by consequent, the varistor is considered the only heat source.

- The thermal link between the varistor and the metal (in our tests is brass) additional mass is very well welded, so heat is changed very fast and uniform between those two structural different pieces. By consequent, both pieces involved will have the same temperature increment, $\Delta\theta_2$.
- This process is considered as fully adiabatic, very fast, and we can correctly apply the energy conservation principle.
- It is obvious that $\Delta\theta_2$ is smaller then $\Delta\theta_1$, as given by (4). This simple additional mass is acting, in fact, like a passive "heat pump", extracting instantaneously a part of the excessive heat stored inside the varistor itself and, by this way, decreasing its temperature and placing the whole assembly inside a possible thermal stability area.

It is mandatory also to explain that, because of the shape and geometry of these supplementary masses, the whole heat dissipation surface is increasing (not as much as in the case of large dedicated radiators). During the permanent service regime, these additional metal masses could be assimilated as radiators (which is not their main role), having a reduced contribution to the heat dissipation balance. By taking in consideration all these ideas, we can reduce, with a few degrees Celsius, the whole varistors' temperature. But, having a reduced additional heat dissipation surface, we cannot consider them as dedicated true radiators.

3. CAD model for the whole structure

The main advantages of placing additional metal masses welded on the varistors could be proved only by using some geometries and numerical models, easily combined and verified by a set of experimental results, in order to perform confirmation of the proposed modeling hypothesis.

All CAD and numerical design procedures were performed at the POLITEHNICA University of Timisoara, Romania. The experimental part was performed at the LAPLACE Laboratory, from the PAUL SABATIER University in Toulouse, France.

All design solutions, simulations and measurements described inside this paper were carried out by using a set of regular 30 mm diameter commercial disk varistors. These devices have reduced 3 mm height and they are applied mainly for a standard 230 V European low voltage (domestic or similar) power supply installations. The varistors involved are not fully covered in epoxy resin, having only the small lateral edge coated with insulator, for about 1 mm. Two Ag alloy based electrodes are deposed by a special solvent technique on both sides. This configuration, having a single varistor alone is called the "A" configuration, in our article.

The additional metallic mass used was a small cylinder made of industrial brass, having a 20 mm diameter and a height of 5 mm. The newly proposed configuration, with a varistor and an additional mass on only one side, centered, is called, in our article, the "B" configuration.

Analysis based on finite elements software is, nowadays, a powerful tool used for modeling heat transfer, as well as electric or magnetic field problems. For a complete CAD modeling, we used the FLUX 2D software, which provided excellent results for pieces having cylindrical symmetry.

The maximum numerically estimated temperature inside that model for configuration A is:

$$\theta_e = \theta_a + \tau_e = 25 + 9.95 = 34.95\ ^\circ C \tag{5}$$

The maximum numerically estimated temperature inside that model, for configuration B is:

$$\theta_e = \theta_a + \tau_e = 26 + 6.38 = 32.38\ ^\circ C \tag{6}$$

In Fig.2 and Fig.3 we will describe the results of the finite elements approach, at 60 s after the overvoltage shock, for each configuration considered:

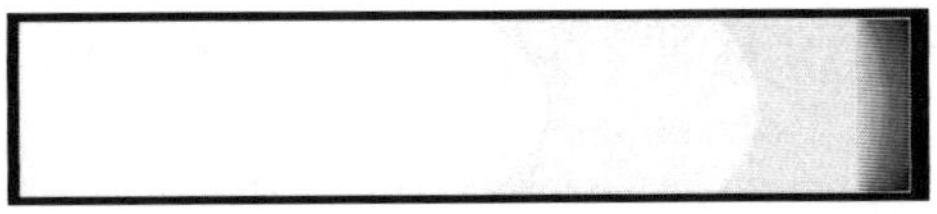

Figure 2. Temperature repartition for configuration A.

Figure 3. Temperature repartition for configuration B.

The maximal temperature obtained for configuration A was 32.84 ˚C and the minimal one was 31.77 ˚C, with a step of 0,09 degrees Celsius for each colour, from yellow to magenta. The time variation graph of the temperature, for a certain random point, located on the top side, where $H = 3$ *mm* and having a $R = 12$ *mm* radius from the axis, belonging to the configuration A, is shown in Fig.4.

The maximal temperature obtained for configuration B, in the same conditions, was around 31.41 ˚C and the minimum was 30.62 ˚C, with a step of 0,07 degrees Celsius for each colour. The time variation of the temperature for that randomly specified point, located on the top side ($H = 3$ *mm*) having a $R = 12$ mm radius, on the configuration B, is shown in Fig.5.

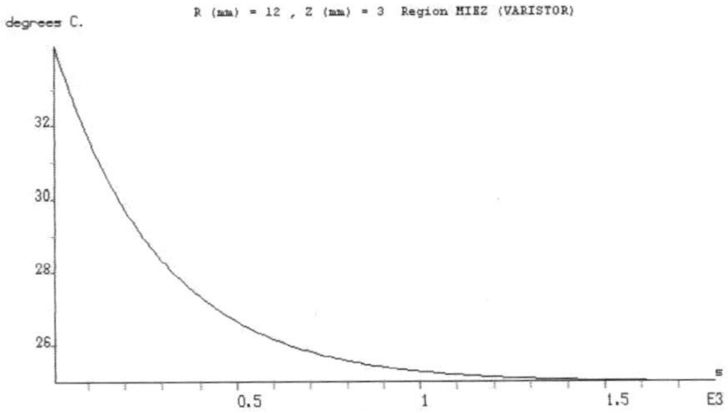

Figure 4. Time variation of temperature for a point on configuration A.

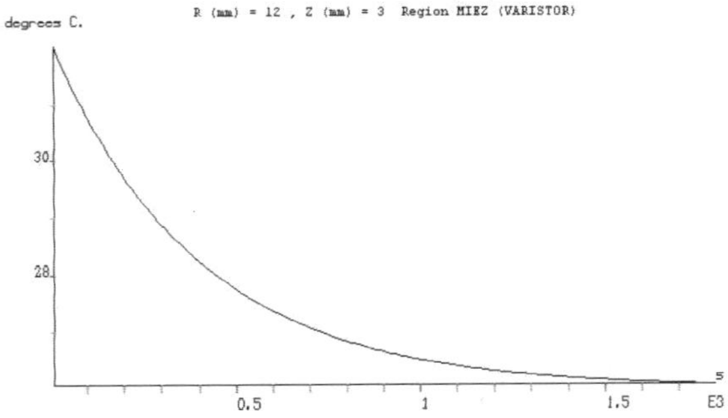

Figure 5. Time variation of temperature for a point on configuration B.

IOP Conf. Series: Journal of Physics: Conf. Series **1195** (2019) 012024 doi:10.1088/1742-6596/1195/1/012024

These graphs are only software estimations for the temperature. They must be verified in practice.

4. Experimental results

Each considered configuration was submitted to an overvoltage test shock, by using a dedicated standard 8/20 shock generator, located at the LAPLACE Laboratory, in Toulouse France.

The maximum real measured temperature θ for the same point, numerically and theoretically considered before (3 mm, 12 mm), belonging to the first configuration A was:

$$\theta = 34{,}44 \text{ °C} \tag{7}$$

We observe that relation (5) offers a very good estimation compared to (7). The time evolution of the temperature, during the "after shock" cooling process, for the same randomly point as before, is shown in Fig.6. The same point as before was also considered for the new configuration B, too.

The maximum real and measured temperature θ, for the same considered point (3 mm, 12 mm), belonging to configuration B was:

$$\theta = 32{,}13 \text{ °C} \tag{8}$$

The cooling process graphs are placed below:

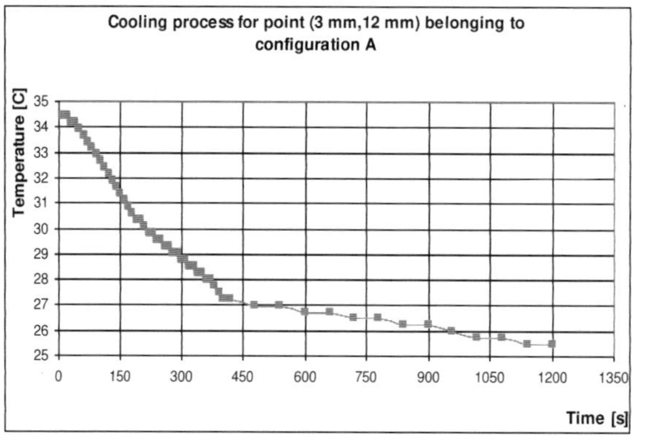

Figure 6. Cooling process time graph for a point located on configuration A.

Equation (6) offers an excellent estimation when compared to (8). The evolution of the temperature during the cooling process, considered for the same point is shown in Fig.7.

The overvoltages applied here are standardized 8/20 μs ones made by a shock generator. All measurements were carried out 60s after applying the shock.

We noticed a very good correspondence between the estimated numerical values and the measured results.

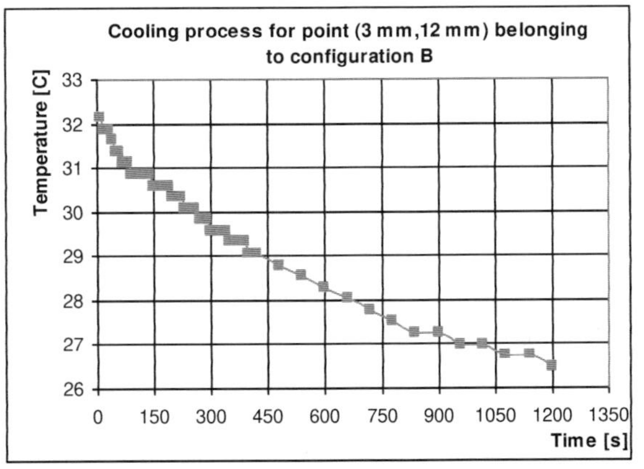

Figure 7. Cooling process time graph for a point located on configuration B.

These are only some simple measurements which are proving the correct measurements on place.

5. Conclusions

After all these studies, we noticed that there is a major risk of overheating a varistor, during permanent or shock (overvoltage) service regime, because the crossing current of that varistor is activated by temperature.

The newly introduced thermal stability concept is an important issue for the quality assessment of a certain varistor. By eliminating any potential overheating risk, the experiments performed in this paper, are placing the varistor inside the thermal stability envelope, in order to avoid its complete destruction and, also, to verify the efficiency of the newly proposed technical solution with additional metal masses.

Potentially, there are plenty of technical solutions concerning the reduction of any overheating risk. From our point of view, the most efficient one involves improvements on the quality of the varistor material itself, which is a study dedicated for chemical or material specialists.

The new and original solution proposed here by the authors, conceived only for existing materials and equipment, consists in an additional metal mass welded on the varistor surface, which acts like a heating pump, extracting heat from the electrical active part of the varistor and relocating this heat in fractions, between the two pieces of equipment.

This original solution is part of our Romanian Patent, RO 117052 B, proved in practice to be efficient, reducing the temperatures compared to the single varistor ones. This technical solution gives us a high energy absorption capability, placing the varistor inside the stability reserve area, for all normal overvoltage shocks.

We observed a very good correspondence between the estimated and the measured results.

There is an almost 2 degrees Celsius difference between the numerical estimated temperature computed by using finite elements model and the real measurements, which is tolerable from the heating point of view.

This simple principle of additional masses could be successfully applied in order to manufacture high energy absorption capacity low voltage varistors or sensors, suitable mostly for low voltage or telecommunications applications.

The use of modern surge protection equipment is mandatory also in the field of any automation equipment, especially in the Power and Energy domain.

References

[1] Entwistle K. M., *Basic Principles of the Finite Element Method*, Alden Press Oxford, Great Britain, 1999
[2] Fakkar A., Bui Ai, Loubiere A., Penchenat G., „Modelisation thermique des varistances à base

d`Oxide de zinc", *Revue générale de l`Electricité,* (in French) Paris, France, no.1, 1993

[3] Tarfulea, N.; Frigura-Iliasa, F. M., Vatau D., Andea Petru, Balcu Florin, Macarie Amalia, "A new Algorithm for the Design of Metal Oxide Varistor Surge Arresters", *Proceedings of the 2016 IEEE 16th International Conference on Environment and Electrical Engineering (EEEIC)*, Florence, Italy, June 7-10, 2016.

[4] Bajenescu T.M., Bazu M., *Component Reliability for Electronic Systems*, Artech House, London, 2009.

[5] Hasse P., "Overvoltage Protection of Low Voltage Systems (*IEEE Power & Energy Series*, 33) Revised Edition", 2001

[6] Frigura-Iliasa F.M., *Contribution à l`étude de l`amélioration de la tenue en énergie des varistances*, Mémoire présenté en vue d`obtenir le diplôme d`ingénieur de l`U.P.T. (in French), Toulouse, France, 1998

[7] Wang, H., a.o., „Nonuniform Heating in Zinc Oxide Varistors Studied by Infrared Imaging and Computer Simulation", *Journal of American Ceramic Society*, Vol. 81, No. 8, August 1999

[8] Nishiwaki, S., Andoh, H., "Failure Mechanisms and Recent Improvements in ZnO Arrester Elements", *IEEE Electrical Insulation Magazine*, Vol. 16, No. 1, January / February 2000

IOP Conf. Series: Journal of Physics: Conf. Series **1195** (2019) 012025 doi:10.1088/1742-6596/1195/1/012025

Optimization of rotation speed for CuSCN hole transport layer in perovskite solar cell using spin coating

J Sulistianto, R W Purnamaningsih, and N R Poespawati*

Department of Electrical Engineering, Universitas Indonesia, Depok 16424, Indonesia

Email: pupu@eng.ui.ac.id

Abstract. Organometal lead halide has gained immense popularity as solar cell active layer due to their unique optical characteristic, low cost, and simple fabrication technique. Transport layer is important component to maximize overall performance of perovskite solar cell. Inorganic CuSCN have good transparency and high hole mobility as p-type transport layer for perovskite solar cell. In this work, we investigated the effect of rotational speed on spin coating to grow CuSCN thin film. Rotational speed is ranging from 1000 rpm to 4000 rpm. The best performing devices is when CuSCN layer deposited at 2000 rpm with V_{OC} 0.96 V, I_{SC} 5.29 mA, and FF 0.48.

1. Introduction

The escalation of world population growth is becoming an attraction for development of new energy sources to meet global energy demand. Photovoltaic based on crystalline silicon is one of the promising devices and has been investigated in past 40 years [1]. Over the last few years, the emerging solar cell such as dye sensitized solar cell (DSSC) and other organic solar cell are also continuously investigated to comply high efficiency and low cost solar energy sources [2]. Perovskite is organic-inorganic material that has crystal structure of ABX_3 and has similar structure with $CaTiO_3$. First organometal halide perovskite as solar cell active layer was performed by Kojima *et al* at 2009 [3]. They used organic-inorganic lead halide perovskite compounds $CH_3NH_3PbX_3$ (X = I, Br) as active layer in similar fabrication method to DSSC. The result was a cell with efficiency 3.81% by using $CH_3NH_3PbI_3$ and TiO_2 as photoanode. These research gained immense popularity and extensively investigated by researcher until 2018, the power conversion efficiency (PCE) of perovskite solar cell can exceed 20% [4]. Such rapid improvement is also caused by low-cost, simple fabrication technique (solution processed or doctor blade), long diffusion length and high absorption coefficient of lead-based perovskite which meet the requirements for mass production and industrial applications [5–7].

From their first appearance, significant progress has been made with various device configurations, including the improvement of transport layer. PEDOT:PSS, spiro-OMETAD, P3HT, and PTAA is organic transport layer and widely used as hole transport layer (HTL). Advantage of using organic material is high performance solar cell can be achieved. However, organic transport layer easily degrades under open environment and accelerates the degradation of perovskite film [8]. In addition to high efficiency, stability of perovskite solar cell must be considered for long term use in applications. Several studies has been conducted and the result mentioned that inorganic material can be used as transport layer. CuSCN [9] and NiO_x [10] were utilized as p-type transport layer, and TiO_2 [11] and

ZnO [12] were used as electron transport layer (ETL). These metal oxides are known in its stability at moisture and posses higher mobility carrier compared to other types.

CuSCN is inorganic p-type which has higher hole mobility (0.01 – 0.1 $cm^2V^{-1}s^{-1}$) and better transparency compared to organic transport layer spiro-OMETAD (4×10^{-5} $cm^2V^{-1}s^{-1}$) [13]. Solution processed in low temperature and simple fabrication method becomes an advantage to apply it in flexible surfaces. There have been several reports about CuSCN as hole transport with different method of fabrication and device configuration. Ye *et al* reported an inverse architecture with PCE 15.6% using electrodeposited method for forming CuSCN thin film [14]. The effect of thickness and annealing temperature of CuSCN has also been studied [15]. Fan *et al* improved the CuSCN surface by modified annealing technique to obtain high quality film [16] and then Xiong *et al* improved CuSCN surface by using PEDOT:PSS to form bilayer HTM [17]. However, it has been observed that reported studies focused on the investigation in inverted configuration and novel technique to improve efficiency instead of n-i-p architecture with simple fabrication method that has been investigated in this paper. Thickness of CuSCN thin film was already studied but not exclusively for optimization of spin coat rotation speed in CuSCN layer. Therefore, in this research we present optimized rotation speed of CuSCN as HTL in perovskite solar cell using spin coating method.

2. Experimental method

2.1. Device structure

In this research, perovskite solar cell was fabricated based on figure 1. FTO coated glass slide was used as front contact and back contact. The device was made on top of fluorine doped tin oxide (FTO) coated glass slide. As an n-type transport layer, mesoporous TiO_2 was used which also works as scaffold for perovskite layer. For the optimization, the rotation speed of spin coat has been varied when deposited CuSCN thin film. Different rotation speed will change the thickness of thin film and control the thickness of CuSCN layer. Furthermore, spin coating is one of an important process to obtain a good thin film.

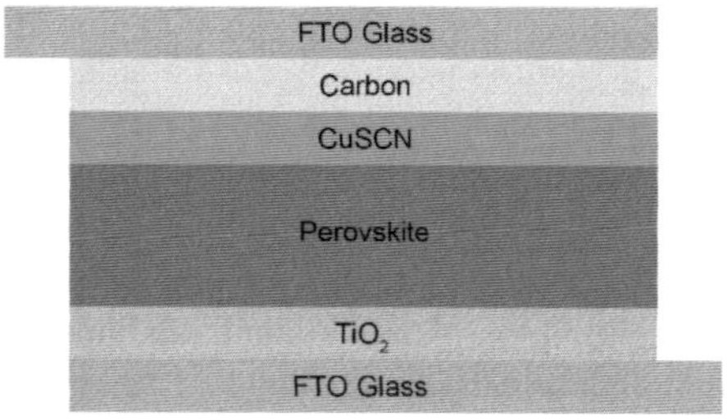

Figure 1. Fabricated perovskite solar cell structure.

TiO_2 was used as electron transport layer because of their ability to collects electron from perovskite into the FTO glass front contact. Due to its wide bandgap (E_g = 3.2 eV), TiO_2 provides great transparency as a thin film, thus it permits more light to perovskite as an active layer. Furthermore, TiO_2 has excellent photocurrent response for visible light when it used together with perovskite [11]. As a hole transport layer in this device, CuSCN was used to collect holes from active layer. It has wide band gap (E_g = 3.6 eV) which has excellent transparency and suitable energy level (E_{VB} = -5.4 eV and E_{CB} = -1.8 eV) for collecting hole [15].

Figure 2 shows energy band diagram of fabricated devices. FTO glass has an energy level -4.4 eV [18]. Energy level for TiO_2, perovskite, and carbon are known from research by Wang *et al* [19]. TiO_2 energy level is -7.3 eV and -4.1 eV for valence and conduction band respectively. Energy level of perovskite is -5.43 eV and -3.93 eV for valence band and conduction band respectively. Carbon has an energy level -5 eV. Perovskite will generate charge when exposes to light and creates electron-hole

pairs. Electron from perovskite conduction band will be transferred to TiO_2 conduction band because of lower energy level in TiO_2 conduction band.

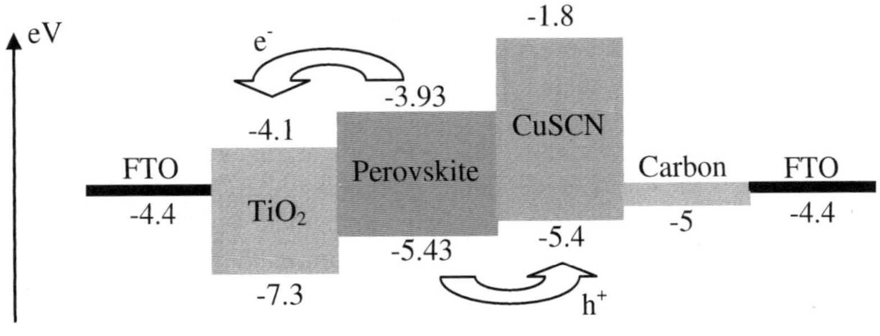

Figure 2. Energy level of perovskite solar cell.

These mechanisms work similarly for hole from perovskite. Hole will move from perovskite valence band to CuSCN valence band due to higher energy level in CuSCN valence band. Carbon was being used to improves performance of hole transfer from HTL to FTO glass.

2.2. Device fabrication

In this research, perovskite solar cell was fabricated on a 2 x 1.5 cm^2 FTO glass. Before fabrication process were conducted, all FTO glass was cleaned using deionized water, acetone, and ethanol, respectively in ultrasonic cleaner bath for 5 minutes each. TiO_2 solution was prepared using 0.2 M titanium (IV) isopropoxide (TTIP) and 0.1 M HCl in anhydrous ethanol. The resultant solutions then stirred for 20 minutes in room temperature. Perovskite solution was prepared from precursor made of methylammonium iodide (MAI) 1.5 M diluted in N, N-dimethylformamide (DMF) and 0.5 M lead(II) chloride. Perovskite precursor then stirred at temperature 80°C for 1 hour before deposited and the solution would become yellow. CuSCN solution was made using 6 mg copper (I) thiocyanate dissolved in 1 ml dipropyl sulphide (DPS) followed by stirring the solution for 8 hours at room temperature.

Fabrication procedure is shown in figure 3. TiO_2 was deposited to FTO glass through spin coating at 2000 rpm for 30 s followed by annealing at 425°C for 30 minutes. Afterward, the sample was being waited until it was cooled down to room temperature. Perovskite solution then deposited above FTO/TiO_2 by spin coating with rotation speed 1000 rpm for 30 s. Sample then being dried at 120°C for 15 minutes and yellow solution of perovskite precursor would turn into grey.

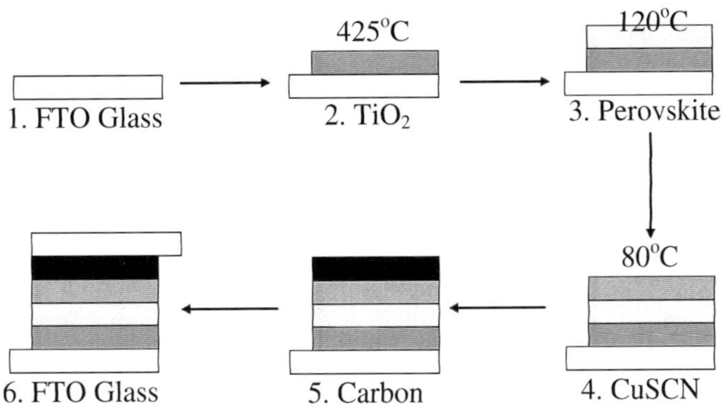

Figure 3. Step by step fabrication of perovskite solar cell.

All samples were fabricated with same configuration except for CuSCN layer which has different rotation speed for each sample. CuSCN was made above perovskite layer by spin coat CuSCN solution at 1000 – 4000 rpm for 30 s followed by annealing at temperature 80°C for 15 minutes. After CuSCN layer was made, carbon powder sprinkled on top of it. Finally, clean FTO glass placed above carbon layer as electrode contact.

3. Result and discussion

Dark I-V curve has been obtained from semiconductor parameter analysis (SPA) HP4145B. Solar cell performance under illumination was performed directly under sun with average irradiance of 800 W/m^2 and measured using multimeter. Photovoltaic parameters of varied CuSCN rotation speed are shown in table 1. Typical I-V curve also is shown in figure 4. All devices perform same fabrication method except in rotational speed for CuSCN layer.

Devices with 1000 rpm and 2000 rpm exhibited better performance in every parameter compared to high speed rotation at 3000 rpm and 4000 rpm. Best result from all devices was obtained by using rotational speed 2000 rpm for CuSCN layer with fill factor (FF) 0.48, short circuit current (I_{SC}) 5.29 mA, and open circuit voltage (V_{OC}) 0.96 V. This result is in line with research that has been conducted by Ye *et al* [14]. It was reported that thin CuSCN film would generate low performance on photovoltaic parameter compared to thicker one. It is important to noting that rotational speed on spin coating has direct consequences to film thickness. Generally, the thickness of spin coated film is proportional to the inverse of the spin speed [20]. This means that low rotation speed will produce thicker film than high rotation speed.

Table 1. Photovoltaic parameters of varied CuSCN rotation speed of perovskite solar cells.

Variation	V_{OC} (V)	I_{SC} (mA)	FF
1000 rpm	0.96	3.82	0.43
2000 rpm	0.96	5.29	0.48
3000 rpm	0.54	1.88	0.37
4000 rpm	0.62	2.99	0.36

Low photovoltaic performance for 3000 rpm and 4000 rpm can be attributed to thickness of CuSCN layer was too thin. CuSCN with thickness lower than 40 nm will produce inhomogeneous and rough surface that lead to inefficient charge transport as mentioned by Jung *et al* [15]. Therefore, smooth CuSCN layer need to be achieved in order to get high performance of perovskite solar cell.

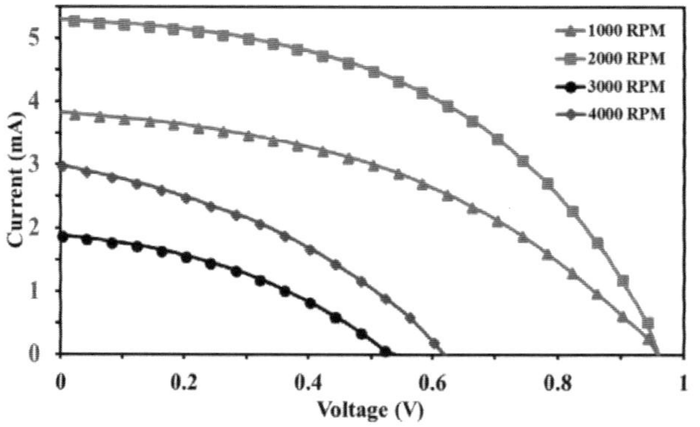

Figure 4. I-V curve of each sample.

The fabricated cell has relatively low performance compared to similar structure demonstrated by Qin *et al* [13]. This might be caused by fabrication process were perform outside clean room or glove box. Moisture and high humidity in not controlled environment can caused perovskite layer degrade. Moisture will decompose $MAPbI_3$ into PbI_2 and make perovskite layer turn back into yellow. Nevertheless, it can take into account that perovskite solar cell is possible to fabricate in such environment.

Inferior in FF might be associated with lack of proper metal contact. Thin film of Au or Ag is usually used as metal contact and deposited with thermal evaporation method. Moreover, it has small difference of energy level to CuSCN. In this research, FTO glass and carbon powder was being used as a contact, which have lower conductivity from mentioned metal contact. Furthermore, enormous difference in energy level from HTL to carbon and FTO back contact might inhibit charge transport. Energy level tailoring is crucial in perovskite solar cell. It can increase V_{OC} and facilitate charge transfer which increase I_{SC} and FF [21].

4. Conclusion

In summary, optimization of rotational speed for CuSCN thin film in perovskite solar cell using spin coating has already achieved. Rotational speed 1000 – 4000 rpm of spin coating was used to deposit CuSCN layer. It was found that devices with high rotational speed such as 3000 rpm and 4000 rpm exhibited low performance. Best devices with V_{OC} 0.96 V, I_{SC} 5.29 mA, and FF 0.48 can be achieved with spin coat rotational speed 2000 rpm. Relatively low photovoltaic parameters might be due to high humidity condition during fabrication. However, this works implies that spin coating rotation speed can be optimized for further application in other layers of perovskite solar cell.

References

[1] Snaith H J 2013 *J. Phys. Chem. Lett.* **4** 3623–30
[2] Li F and Liu M 2017 *J. Mater. Chem. A* **5** 15447–59
[3] Akihiro Kojima, Teshima K, Shirai Y and Miyasaka T 2009 *J Am Chem Soc* **131** 6050–1
[4] Mesquita I, Andrade L and Mendes A 2018 *Renew. Sustain. Energy Rev.* **82** 2471–89
[5] Burschka J, Pellet N, Moon S J, Humphry-Baker R, Gao P, Nazeeruddin M K and Grätzel M 2013 *Nature* **499** 316–9
[6] Lin Q, Armin A, Nagiri R C R, Burn P L and Meredith P 2015 *Nat. Photonics* **9** 106–12
[7] Wehrenfennig C, Liu M, Snaith H J, Johnston M B and Herz L M 2014 *Energy Environ. Sci.* **7** 2269–75
[8] Habisreutinger S N, McMeekin D P, Snaith H J and Nicholas R J 2016 *APL Mater.* **4**
[9] Arora N, Dar M I, Hinderhofer A, Pellet N, Schreiber F, Zakeeruddin S M and Grätzel M 2017 *Science.* **358** 768–71
[10] Jiang F, Choy W C H, Li X, Zhang D and Cheng J 2015 *Adv. Mater.* **27** 2930–7
[11] Etgar L, Gao P, Xue Z, Peng Q, Chandiran A K, Liu B, Nazeeruddin M K and Grätzel M 2012 *J. Am. Chem. Soc.* **134** 17396–9
[12] Farrag A A G and Balboul M R 2017 *J. Sol-Gel Sci. Technol.* **82** 269–79
[13] Qin P, Tanaka S, Ito S, Tetreault N, Manabe K, Nishino H, Nazeeruddin M K and Grätzel M 2014 *Nat. Commun.* **5** 1–6
[14] Ye S, Sun W, Li Y, Yan W, Peng H, Bian Z, Liu Z and Huang C 2015 *Nano Lett.* **15** 3723–8
[15] Jung J W, Chueh C C and Jen A K Y 2015 *Adv. Energy Mater.* **5** 1–7
[16] Fan L, Li Y, Yao X, Ding Y, Zhao S, Shi B, Wei C, Zhang D, Li B, Wang G, Zhao Y and Zhang X 2018 *ACS Appl. Energy Mater.* **1** 1575–1584
[17] Xiong Q, Tian H, Zhang J, Han L, Lu C, Shen B, Zhang Y, Zheng Y, Lu C, Zeng Z, Hu Z, Wu L and Zhu Y 2018 *Org. Electron.* **61** 151–6
[18] Akkerman Q A, Gandini M, Di Stasio F, Rastogi P, Palazon F, Bertoni G, Ball J M, Prato M, Petrozza A and Manna L 2017 *Nat. Energy* **2**
[19] Wang H, Hu X and Chen H 2015 *RSC Adv.* **5** 30192–6

[20] Chaki S H, Joshi H J, Tailor J P and Deshpande M P 2017 *Mater. Res. Express* **4** 076402
[21] Kim H, Lim K-G and Lee T-W 2016 *Energy Environ. Sci.* **9** 12–30

Acknowledgments
This work has been supported by International publications indexed for the Final Project of Students (PITTA) Research Grant 2018 from Universitas Indonesia with Nr. 2502/UN2.R3.1/HKP.05.00/2018.

IOP Conf. Series: Journal of Physics: Conf. Series **1195** (2019) 012026 doi:10.1088/1742-6596/1195/1/012026

Fuzzy Tuning and Power Reaching Law-Based Discrete Sliding Mode Control for Solar Photovoltaic Inverters

En-Chih Chang, Hung-Liang Cheng and Chun-An Cheng

Department of Electrical Engineering, I-Shou University, Taiwan, R.O.C.
E-mail: enchihchang@isu.edu.tw

Abstract. This paper develops a digital signal processor (DSP)-controlled solar photovoltaic inverter that is well realized to prove the proposed controller. The discrete sliding mode control (DSMC) with power reaching law (PRL) not only suits for digital implementation because of its finite sampling frequency but permits faster convergence speed. However, the chattering around PRL-based DSMC exists and may cause excessive power losses. A fuzzy tuning technology is employed to handle the problem of the chattering so that the robustness of the system can be increased. Simulation results display that the proposed solar photovoltaic inverter can generate low total harmonic distortion (THD) under rectifier-type load circumstances and fast transient response under step load changes. Experimental results on a solar photovoltaic inverter laboratory prototype controlled by a digital controller are provided to substantiate both theoretical analysis and simulation results.

1. Introduction

With the raising indispensability for high-quality AC power supplies, solar photovoltaic inverters have been adopted as an important part for sustainable energy systems [1], [2]. Solar photovoltaic inverters are demanded to keep high performance sine output voltage of low harmonic distortion and fast transience even feeding highly nonlinear loads that can be achieved by employing feedback controller. With the availability of high-frequency switching devices and the popularity of the DSP, numerous digital control methods have been applied in solar photovoltaic inverters [3-5] and one of them is sliding mode control (SMC) because of its insensitivity against system uncertainties [6-8]. The SMC methodology is initially developed from a continuous-time perspective. If it is implemented on a digital computer, the limited sampling rate, sample/hold effects and discrete errors will worsen system performance. Thereby we present a power reaching law (PRL)-based discrete sliding mode control (DSMC) to carry out digital implementation and simultaneously accelerates convergence rate in the design of solar photovoltaic inverters [9-12]. But, high frequency oscillations called chattering that may lead to the excitation of unmodeled dynamics and even damage the plant. A fuzzy tuning technology has been effectively used in various science and engineering fields and possesses many advantages, such as model-free, universal approximation theorem and rules-based algorithm. It relies on the human capability to understand the behaviour of the system, and provides qualitative control rules which do neither request an exact mathematical modelling of the system nor complicated computations [13-15]. The FL is thus incorporated in PRL-based DSMC to eliminate the effect of the chattering. By using the association of the FL and PRL-based DSMC, simulations and experiments on a solar photovoltaic inverter prototype have proved the achievement of very low THD under nonlinear loading and fast dynamic response under transient loading.

IOP Conf. Series: Journal of Physics: Conf. Series **1195** (2019) 012026 doi:10.1088/1742-6596/1195/1/012026

2. System Modelling and Controller Design

Fig. 1 depicts a generally utilized solar photovoltaic inverter, which is composed of an LC filter and switching component with MOSFET transistors. By applying KVL and KCL, the state space equations of solar photovoltaic inverter can be formulated as

$$L\frac{di_L}{dt} + v_c = K_{pwm}v_{con} \tag{1}$$

$$i_L = C\frac{dv_c}{dt} + \frac{v_c}{R} \tag{2}$$

where $K_{pwm} = V_d / \hat{v}_{tri}$ the equivalent gain of the inverter, V_d = the dc supply voltage, $\hat{v}_{tri}$ = the amplitude of the triangular wave of the PWM, and v_{con} = the control signal.

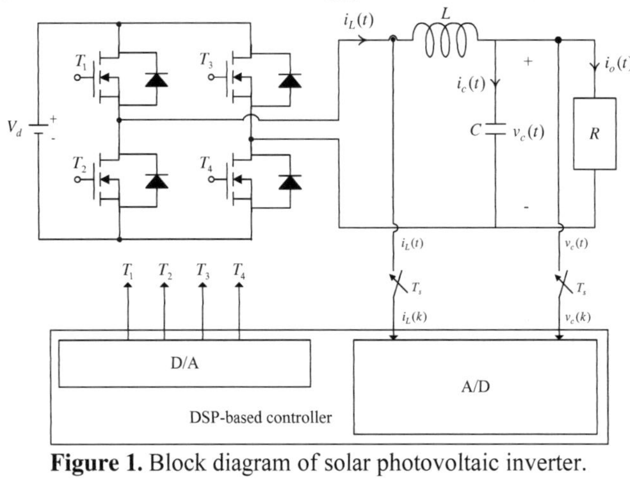

Figure 1. Block diagram of solar photovoltaic inverter.

Substituting (2) into (1), the differential equation is expressed as follows.

$$\ddot{v}_c = -\frac{1}{RC}\dot{v}_c - \frac{1}{LC}v_c - \frac{K_{pwm}}{LC}v_{con} \tag{3}$$

Defining $x_z = [v_c, \dot{v}_c]^T$, and $u = v_{con}$, the (3) can be rewritten as

$$\begin{bmatrix}\dot{x}_{z1}\\ \dot{x}_{z2}\end{bmatrix} = \underbrace{\begin{bmatrix}0 & 1\\ -\frac{1}{LC} & -\frac{1}{RC}\end{bmatrix}}_{A}\begin{bmatrix}x_{z1}\\ x_{z2}\end{bmatrix} + \underbrace{\begin{bmatrix}0\\ \frac{K_{pwm}}{LC}\end{bmatrix}}_{B}u \tag{4}$$

The fuzzy SMC with PRL is now employed to make controlled solar photovoltaic inverter resist the prescribed load uncertainty while keeping low distorted output-voltage and achieving robust performance.

For digital implementation purpose, the discrete-time equation can be stated as

$$x_z(k+1) = \Phi x_z(k) + \Gamma u(k) + w(k) \tag{5}$$

where $x_z(k) = [v_c(k) \quad \dot{v}_c(k)]^T$, $\Phi = e^{AT_s}$, $\Gamma = \left(\int_0^{T_s} e^{A\tau}d\tau\right)B$, and T_s is the sampling period, $w(k) = i_o(k)$.

In order to enforce the output $x_z(k)$ to $x_{ref}(k)$. Therefore, a tracking error $x_{ze}(k)$ can be expressed as

$$x_{ze}(k) = x_z(k) - x_{ref}(k) \tag{6}$$

where $x_{ref}(k)$ stands for reference sinusoidal voltage.

From (5) and (6), the error dynamics yields

$$x_{ze}(k+1) = \Phi x_{ze}(k) + \Gamma u(k) + \psi(k) \tag{7}$$

where $\psi(k) = \Phi x_{ref}(k) - x_{ref}(k+1) + w(k)$.

The following discrete sliding surface $s(k)$ can be constructed by

$$s(k) = c x_{ze}(k) \tag{8}$$

where c is a matrix.

Then,

$$s(k+1) = c x_{ze}(k+1) = c\Phi x_{ze}(k) + c\Gamma u(k) + c\Phi x_{ref}(k) - c x_{ref}(k+1) + c w(k) \tag{9}$$

The power reaching law can be introduced as

$$s(k+1) = s(k) - \varepsilon T_s |s(k)|^{\gamma} \operatorname{sgn}(s(k)) \tag{10}$$

where $0 < \varepsilon T_s < 1$ and $0 < \gamma < 1$.

From (9) and (10), we have

$$u(k) = -(c\Gamma)^{-1}\left[c\Phi x_{ze}(k) + c\Phi x_{ref}(k) - c x_{ref}(k+1) - s(k) + \varepsilon T_s |s(k)|^{\gamma} \operatorname{sgn}(s(k)) + c w(k)\right] \tag{11}$$

The sgn function is shown in (11) and it frequently has the chattering in practice. Thus, the sgn function and sliding surface are fuzzified, so that the chattering can be eliminated. The sliding surface is a particular fuzzy case and its membership function can be defined below.

$$\Lambda_{s=0} = \begin{cases} 1 & , \quad s = 0 \\ 0 & , \quad \text{otherwise} \end{cases} \tag{12}$$

Then, the human experience is fixed as linguistic rules in the following.

IF far apart from the sliding surface THEN increase feedback gain (13)

3. Simulation and Experimental Results

The parameters of the solar photovoltaic inverter used for the proposed controller are listed in Table 1. In order to test the performance of the transience further, Fig. 2 (a) show the simulated output voltage and the load current obtained using the proposed controller under step load change from no load to full load at a 90 degree firing angle. As can be seen, the proposed controller shows slight instant voltage droop and fast recovery of the steady-state response. Similarly, the transience with linear resistive load is explored and Fig. 2(b) displays the experimental result with step load change from no load to full load at a 90 degree firing angle. The transient behaviour is satisfactory and the output-voltage droop with fast retrieval is also tiny. Figure 3(a) show the simulated output voltage and the load current waveforms of the solar photovoltaic inverter controlled by the proposed controller under rectifier load (parallel resistor, 40 Ω and capacitor filter, 200μF). A very low output-voltage distortion (%THD is 0.91%) appears at the output voltage while the current ascends suddenly. The experimental performance of the proposed controller with rectifier load (capacitive filter 200μF and 40 Ω resistive load) is reported, too. Fig. 3(b) represents that the experimental output voltage is almost sine wave and produces low distortion (%THD of output-voltage is 1.92%), which exceeds the industrial standard. In the phase planes, the classic SMC shown in Fig. 4(a) with a chatter vibration converges to the origin. The transient response and steady-state errors are unsatisfactory. However, the state trajectory of the proposed controller plotted in Fig. 4(b) provides fast and smooth convergence to zero. Thereby, the proposed controller does effectively eliminate the chattering and decrease steady-state errors.

Table 1. System parameters

DC-link voltage	V_d= 200 V
Filter inductor	L = 0.2 mH
Filter capacitor	C = 5 μF
Resistive load	R=12 Ω
Output voltage and frequency	v_c = 110 V_{rms}, f = 60 Hz
Switching frequency	f_s = 18 kHz

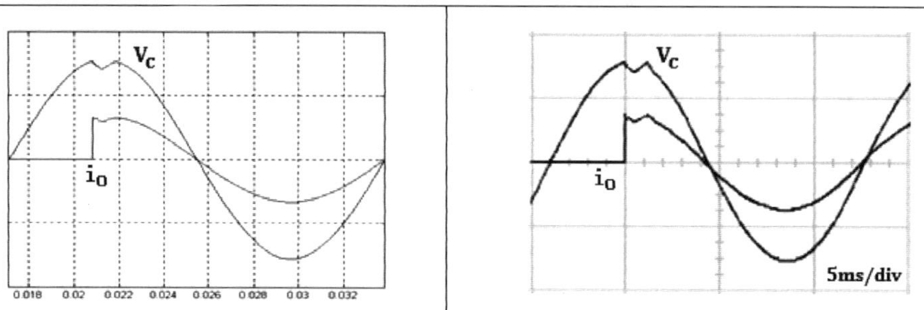

(a) Simulation (b) Experiment

Figure 2. Proposed controller under step load changes (100 V/div; 20 A/div; 5 ms/div).

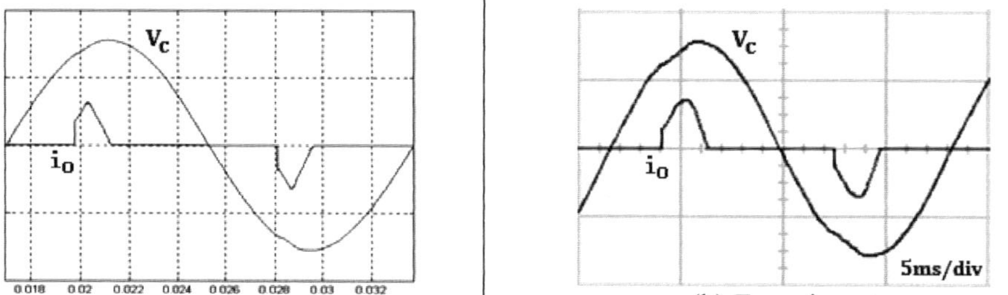

(a) Simulation (b) Experiment

Figure 3. Proposed controller rectifier loads (100 V/div; 30 A/div; 5 ms/div).

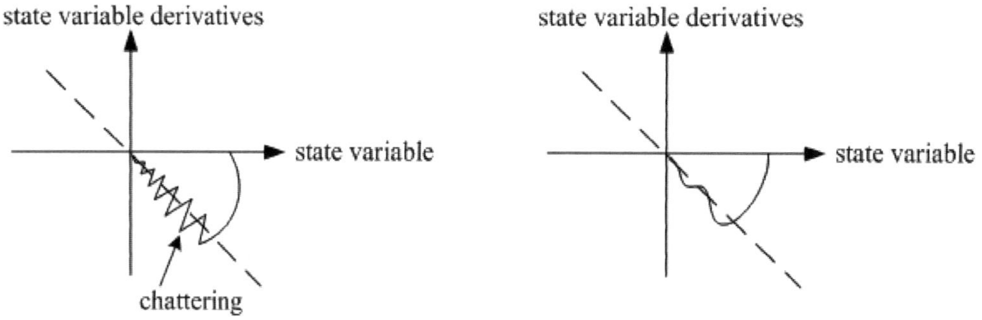

(a) (b)

Figure 4. Phase planes (a) Classic SMC with chattering (b) Proposed controller without chattering.

4. Conclusions

A combination of the Fuzzy tuning technology and the PRL-based DSMC for solar photovoltaic inverters is proposed to maintain low distorted output-voltage even under non-linear loading. As compared to continuous-time SMC, the DSMC can speed up the convergence rate and is more suitable for digital implementation, while the Fuzzy tuning devotes to eliminate the chatter phenomenon. Experimental and simulation results corroborate the applicability and efficacy of the proposed controller.

Acknowledgment

This work was supported by the Ministry of Science and Technology of Taiwan, R.O.C., under contract number MOST 107-2221-E-214-006.

IOP Conf. Series: Journal of Physics: Conf. Series **1195** (2019) 012026 doi:10.1088/1742-6596/1195/1/012026

References

[1] P. Breeze, Solar Power Generation, Academic Press, 2016.

[2] R. Teodorescu, M. Liserre, and P. Rodríguez, Grid Converters for Photovoltaic and Wind Power Systems, Wiley, New York, USA, 2010.

[3] M. Gangavarapu, "Perturb and Observe MPPT Algorithm Implementation for PV Applications," International Journal of Computer Science and Information Technologies, vol. 6, no. 2, pp. 1884-1887, 2015.

[4] I. P. Ratna, W. Sapto, and R. Muhamad, "Maximum Power Point Tracking for Photovoltaic Using Incremental Conductance Method," Energy Procedia, vol. 68, pp. 22-30, 2015.

[5] R. J. Wai, C. Y. Lin, Y. C. Huang, and Y. R. Chang, "Design of High-Performance Stand-Alone and Grid-Connected Inverter for Distributed Generation Applications," *IEEE Trans. Ind. Electron.*, vol. 60, no. 4, pp. 1542-1555, 2013.

[6] X. Hao, X. Yang, T. Liu, L. Huang, and W. J. Chen, "A Sliding-Mode Controller With Multiresonant Sliding Surface for Single-Phase Grid-Connected VSI With an LCL Filter," *IEEE Trans. Power Electron.*, vol. 28, no. 5, pp. 2259-2268, 2013.

[7] A. Abrishamifar, A. A. Ahmad, and M. Mohamadian, "Fixed Switching Frequency Sliding Mode Control for Single-Phase Unipolar Inverters," *IEEE Trans. Power Electron.*, vol. 27, no. 5, pp. 2507-2514, 2012.

[8] R. Aghatehrani, and R. Kavasseri, "Sensitivity-Analysis-Based Sliding Mode Control for Voltage Regulation in Microgrids," *IEEE Trans. on Sustainable Energy*, vol. 4, no. 1, pp. 50-57, 2013.

[9] W. Gao and J. C. Hung, "Variable structure control of nonlinear systems: A new approach," *IEEE transactions on Industrial Electronics*, vol. 40, no. 1, pp. 45–5, February 1993.

[10] V. K. Thakar, "Power rate reaching law based discrete sliding mode control," *Annual Progress Seminar 2005.*

[11] W. Gao, Y. Wang, and A. Homaifa, "Discrete-time variable structure control systems," *IEEE transactions on Industrial Electronics*, vol. 42, no. 2, pp. 117–122, April 1995.

[12] S. C. Tan, Y. M. Lai, and C. K. Tse, Sliding Mode Control of Switching Power Converters: Techniques and Implementation, CRC Press, *Boca Raton*, *FL*, USA, 2012.

[13] X. S. Yang, Nature-Inspired Algorithms and Applied Optimization, Springer, New York, 2018.

[14] Y. Y. Tzou, and R. S. Ou, "DSP-Based Feedforward Fuzzy Control of a PWM Inverter for AC Voltage Regulation," *Journal of Control Systems and Technology*, vol. 2, No. 4, pp. 231-238, 1994.

[15] A. Syropoulos, *Theory of Fuzzy Computation*, Springer-Verlag, New York, 2014.

AUTHOR INDEX